Our Cover Girl is Claire Bouskill, from Leeds, England. She currently studies criminology at Teeside University in Middlesborough, England.

STANDING TALL

WITH

TURNER SYNDROME

Essays by Women with Turner Syndrome

Xo

Edited by
Claudette Beit-Aharon

Photography, Layout, and Typesetting by
Brigitt Angst

Medical Consultant
Lynne L. Levitsky MD, Chief of Pediatric
Endocrinology, Massachusetts General
Hospital, Boston

Copyright © 2013 by Claudette Beit-Aharon

All rights reserved. No part of this book may be used or reproduced or transmitted in any manner whatsoever or by any means, electronic or mechanical, including photocopying, recording or by any information storage and retrieval system without written permission except in the case of brief quotations embodied in articles and reviews.

Although the editor, authors and publisher have made every effort to insure accuracy and completeness of information contained in this book we assume no responsibility for errors, inaccuracies, omissions or inconsistencies herein. Any slights of people, places or organizations are unintentional.

Published by Nanomir Press
Newton, MA, USA, 2013
stwts18@gmail.com

ISBN 978-1-291-50771-3

Library of Congress Control Number:
2013916917

First Edition October 24, 2013

This book is dedicated to all those exploring the impact of Turner syndrome.

ACKNOWLEDGEMENTS

I want to express my heartfelt thanks to Lynne L. Levitsky MD, chief of the Pediatric Endocrine Unit at the Massachusetts General Hospital for Children in Boston. Her unwavering support and input has helped make this work possible. Jonathan Rhodes, MD, pediatric cardiologist at Children's Hospital in Boston, reviewed the cardiology portions and explained the intricacies of cardiac development to me in a way that removed much of the mystery, enabling me to share what I learned. Angela Lin, MD, staff geneticist at Massachusetts General Hospital, clinical professor of pediatrics at Harvard Medical School and associate editor of the American Journal of Medical Genetics, reviewed the genetics portions and ensured accuracy, as well as opening my eyes to new possibilities.

I am also grateful for the support and encouragement of Sharon Terry, the president and CEO of the Genetic Alliance, a network of disease advocacy organizations enabling individuals, families and communities to reclaim their health and become full participants in translational research and services.

TABLE OF CONTENTS

FOREWORD BY LYNNE L. LEVITSKY, MD   xiii

EDITOR'S INTRODUCTION   xvii

CONTRIBUTORS   xxiii

ESSAYS

The Cat's Game	Fiona Wilson	1
The Unmentionable Condition	Joy Webster	5
View from the Yellow Brick Road	Wendy Coates	11
Hitting the Curve Ball	Caroline Keras	15
Passing the Torch	Michelle Orofino	19
Expect the Unexpected	Camille Gaddis	27
Let's Talk about Sex	Diana Clifton	35
Fragments of Me, Photo Essay	Brigitt Angst	43
Embracing Scientific Inspiration and Faith through TS	Rebecca Green	51
Peter Pan and Me	Miriam Beit-Aharon	57
Who Am I?	Caroline Skene	63
Mountain Ways	Nadine Chaluck	69
Rejecting Denial and Finding Myself	Jennifer Liu-Mormile	75
Damaged Goods or Specialty Item	Jessica Esau	83

A Song in My Heart	Patricia Ann Selway	87
Growing Up, I Never Felt Different	Stefania Dimaio	93
Changes	Jordan Brilhante	97
The Circle of My Life	Susan Lazar	101

APPENDICES

Appendix 1:

Discussion Topics by Susan Lazar, Psychotherapist

Adult Women Discussion Group	109
Teens Discussion Group	111
Parents Discussion Group	113

Appendix 2:

Karyotyping in Plain English	117

Appendix 3:

Hormones and Some Facts about Our Favorites	129

Appendix 4:

Turner Syndrome: Frequently Asked Questions	137

Appendix 5:

Recommended Reading	155

Appendix 6:

Recommended Viewing	159

Appendix 7:

Support Organizations   163

Appendix 8:

Glossary   171

Foreword
Lynne L. Levitsky, MD

As a pediatric endocrinologist, my work with Turner syndrome focuses on girls and issues of puberty and growth, as well as the implications of the diagnosis, which is new to girls and their families. Because they graduate from pediatric care, I have rarely been privy to their lives after young adulthood.

These essays, written by women of different ages who are determined and strong, are a revelation. The younger women have a sense of purpose and enthusiasm for the future, and the older women are living up to their individual goals and demonstrating resilience and flexibility. They form lives of their own making often defying the expectations of others. These women help prove, that the renowned lawyer and Congresswoman Bella Abzug was right when she said that the test for whether or not you can hold a job should not be the arrangement of your chromosomes.

The personal narratives are bound together by an excellent description of Turner syndrome, the chromosomal rearrangements associated with it, the diverse physical ramifications that may result, and present treatment options. It will prove invaluable to girls and their families and ease their struggle to understand the diagnosis and its implications.

Women and girls with Turner syndrome will learn from this book that good and comprehensive medical care can anticipate diverse problems and ameliorate severe long-term consequences. Parents of young children with Turner syndrome will understand that the spectrum of the condition is so wide that it is hard to predict individual outcomes. Medical professionals should become more aware through this book that their patients are women of normal intellect with a capacity for humor, creativity, empathy, and personal growth.

Language used in the past (and sometimes still today), included the term "feminization" rather than "induction of puberty," thus falsely giving young girls a sense that they were not female enough. At the same time, physicians' focus on physical changes like growth and puberty may mean that young women do not receive adequate care and evaluation of their special learning needs, seen even in very intelligent young women with Turner syndrome.

Transition from pediatric to adult care can bring a new set of concerns, as internists must focus on continued monitoring for bone density problems, carbohydrate and lipid metabolism issues, and the possible development of autoimmune disorders.

Being special is not always an asset when it comes to medical care, but in most respects girls and women with Turner syndrome are less special than they might think. Many other people with an unremarkable chromosomal makeup have disorders that are also associated with Turner syndrome, such as high blood pressure, hypothyroidism, or heart disease. The treatments are not different.

Participation in civic advocacy communities has benefited women with Turner syndrome. I invoked the power of the national Turner Syndrome Society and the National Organization for Women (NOW) to help gain insurance coverage for growth hormone therapy in Turner syndrome by a major health insurer in the United States. I mentioned that height in Turner syndrome is a feminist issue and made sure that my letter to the insurer was also sent to the Turner Syndrome Society and NOW. This mysteriously lead to the immediate approval of growth hormone therapy by the reluctant insurer.

The power of peer relationships described in these essays is unique. So many women never met anyone else with Turner syndrome as they were growing up. Some come from families where secrecy is an important part of their life experience. Therefore medical caregivers need to provide the information to facilitate participation in social activities with peers who have the same condition. Parents need to understand the importance of these relationships for their daughters. The freedom and relief that comes from being open and meeting others is obvious in many of these life stories. Sharing information on where to find flattering clothing and shoes in small sizes, in addition to coping with more serious complications of the condition is made easier if national directories of resources are available. Many organizations in the United States, Canada and the United Kingdom are listed in this book.

Sadness due to infertility is another important aspect of these essays. Whether assisted fertility techniques will be appropriate, given the possible risk of cardiac problems is presently unclear. Family members, friends, and physicians need to be good and supportive listeners when discussing alternative means of building families through adoption, marriage, and strong friendships. The world is a much more open place than it was when some of these women were young adults, and these possibilities are now more easily discussed and realized.

Many of the essayists describe wonderful, warm, and compassionate care by their physicians, but there were enough references to callous and inappropriate medical treatment to make me wince. Families and women should not suffer silently. Insist on respectful and knowledgeable care by health professionals. It is your right. Lending your doctor this book may be a good place to start.

The insight that these writers have given me has made me a more empathic physician and a better human being. I strongly recommend it to all physicians and health care providers who are involved in any aspect of care for girls and women with Turner Syndrome.

Boston, September 2013

Editor's Introduction

Standing Tall with Turner Syndrome is designed to speak to three groups: women and girls with Turner syndrome, their families, and clinicians. Women who have Turner syndrome wrote all of the essays, and have approved my editorial contributions. Medical experts in endocrinology, cardiology, and genetics have also reviewed the content.

The genesis of this book was in the winter of 2009, when my daughter wrote her college admissions essay on the subject of going through puberty with Turner syndrome. An expanded version of that essay, *Peter Pan and Me,* is included in this collection. My daughter shared her writing with six friends from Turner syndrome camp who were visiting us for a week, prompting a discussion about how they felt misrepresented in popular and even medical literature and how too many people who should know better, including some doctors and other medical personnel, were misinformed about them, their needs, and their life experiences. They wanted to contribute in some way to a solution, and essays by three of these young women appear here.

This book focuses on the essayists' perspectives on their lives as girls and women, which includes each writer's experience of having Turner syndrome. The aim is to encourage girls with Turner syndrome to grow into happy and confident women, while feeling the support of others who truly understand their dreams and challenges. Also we hope that parents, siblings, spouses, and extended family will feel better equipped to support their loved one with Turner syndrome, and that clinicians in particular might find insights by reading these first person accounts.

Our contributors cover a wide spectrum of the Turner syndrome experience and range in age at the time of the writing from late teens to mid-fifties. Young women and their families can look to our essayists to see how adults with Turner syndrome have built fulfilling lives and careers in fields such as accounting, cell biology, computer science, education, library science, psychotherapy, neurobiology, social work, massage therapy, theatre, pharmacology, photography, and professional writing. One is currently a crew member on a tall ship. Four are married, and one is a widow.

Our writers come from Canada (three) and the United Kingdom (four) as well as the United States. The Americans hail from Arizona, Kansas, Massachusetts, New Jersey, New York, North Carolina, and Texas and represent varied ethnic backgrounds. More than half are bilingual, speaking languages in addition to English including Chinese, Swiss German, Hebrew, French, Italian, and Spanish.

Turner syndrome is most often depicted only as a collection of maladies and malformations. I have often heard women or girls with Turner syndrome recite this litany of potential symptoms, as if by rote, when asked to describe the condition. The lists of problems found on medical and organizational websites, brochures, and other resources are long and terrifying. They are sometimes accompanied by statistical percentages suggesting how common a particular feature might be, but they do not clearly explain that a given person will never have all of the possible manifestations, or even most of them. We are grateful that a young woman today has access to medical advances not yet available when her older sisters were growing up, but there is still a long way to go. Too many girls are not diagnosed at an early age, when they could take full advantage of all the treatments available that improve both short and long-term health. Often, the psychological and emotional issues that come with a Turner syndrome diagnosis are neglected in favor of a singular focus on medical or physiological issues.

Doctors, families, and the women themselves need to know what to look out for after diagnosis. But they need information on counseling. This information is somewhat more available than it was only a few years ago.

As of this writing, online excerpts from medical textbooks are still replete with outdated and offensive pictures of women and girls, naked, their eyes covered by a black rectangle. These photographs are not only upsetting to parents, prospective parents, and women with Turner syndrome but can unintentionally lead to dehumanizing treatment from the medical community. This book is here to say that Turner syndrome is not incompatible with a satisfying and accomplished life. A woman carrying a fetus diagnosed with Turner syndrome should not despair about her daughter's capacity for happiness.

Many of the individual symptoms that, when combined, can be present in those with Turner syndrome also occur in the general population. Women and girls with Turner syndrome can reap the benefits of well-tested therapies and join non-Turner medical communities for information and support, whether coping with infertility or other medical issues. Premature ovarian failure, hearing loss, coarctation of the aorta, bicuspid aortic valve, short stature, low thyroid, and diabetes occur commonly in the general population for different reasons.[1]

Readers who are parents and family members will find information here about what their loved one with Turner syndrome faces. A cardiologist, nephrologist, obstetrician/gynecologist, endocrinologist, or geneticist who has little or no experience treating someone with Turner syndrome can become more aware of the facts and concerns of these women. We aspire to help clinicians see and treat each woman

---

[1] According to the Centers for Disease Control and Prevention (CDC), infertility, for example, affects twelve percent of the reproductive age population in the United States. (See www.cdc.gov/ART/)

as the whole person she is. Through these essays, we hope medical professionals will feel better equipped to meet a woman who has Turner syndrome on her own terms and thereby foster more positive doctor-patient relationships.

It is often hard for parents to guess what a daughter's issues will be, or what she feels are her biggest problems. The child may not share the same concerns that the parents find overwhelming, and vice versa. For example, while attending her first Turner syndrome conference, one girl said that meeting a woman in her sixties was the most exciting and important part. Until then she had heard talk only of "Turner girls" and never about mature women. With that flawed information, she had logically concluded that there were no adult women with Turner syndrome, and that she would die before becoming a woman at all.

Having access to the varied experiences of the essayists in this book may help those with Turner syndrome and their families avoid the turmoil engendered by such confusion and misinformation. A woman or girl diagnosed early will know about the likelihood of infertility in advance and can prepare for what will be necessary if she chooses to become a parent in the future. But the profound sense of loss comes far earlier than for many other infertile women. Extra emotional support may be required when processing this aspect of the diagnosis.

Whenever a child has a potentially serious medical condition, it is natural for others, particularly parents, to be overprotective. Expectations and attitudes towards the "cute" or "small package" can create obstacles to maturity. As a young woman makes her way in the world of work and relationships, she should see herself as a person who deserves to be taken seriously by family, coworkers, and physicians and celebrate her joys as well as face her challenges.

As with many conditions that affect women, people in supportive roles often respond with an overabundance of teddy bears, ribbons, and omnipresent pink. For girls and women who have Turner syndrome and delayed puberty, this can unintentionally encourage a childish self-image beyond a time when that is appropriate. As several of the essays that follow illustrate, women and girls with Turner syndrome often struggle to be treated in an age appropriate manner. A person can get worn down by societal and or medical expectations, and develop a habit of seeing themselves as less mature than their peers. Over time this can be destructive to a sense of competence, confidence, or ability to develop into a full-fledged adult.

This book attests to the strength, accomplishments, and wisdom of the essayists. The authors send their love to women with Turner syndrome everywhere in the hope that they will feel a sense of shared experience and community support.

Claudette Beit-Aharon
Newton, Massachusetts, October 2013

Contributors

The names in quotation marks are pseudonyms for professional reasons.

Brigitt Angst was born in 1958 and grew up in Lucerne, Switzerland. She has a PhD in clinical immunolgy and worked as a researcher in cell and molecular biology for twenty years. She lives in London, England. More recently she studied photography, obtaining an MA with Distinction from De Montfort University in 2011. Presently she works as a wellness entrepreneur and photographer. Her photography projects often focus on collaborative portraiture including projects on TS and this book. She relishes the input of her models. She has practiced Iyengar yoga since 1988.

Miriam Beit-Aharon was born in 1991 and is from Newton, Massachusetts. She graduated from the University of Massachusetts in Amherst in 2013 with a degree in Understanding and Documenting Maritime Culture and a minor in anthropology. Her first love is the ocean and sailing on tall ships, and at the time of this writing she is crewing on the schooner *Mystic*. She enjoys singing a cappella, dancing, and origami. She is grateful for her family, her TS sisters, and friends—whose company she cherishes.

Jordan Brilhante was born in 1991 and is from Phoenix, Arizona. At Mesa Community College she majored in communications, transferring to Arizona State University in the fall of 2012. She still works on her relationships everyday, and cherishes the beautiful life with which she is blessed.

Nadine Chaluck was born in 1996 and calls Golden, British Columbia home. She attended Golden Secondary School, and wrote this essay when she was still in high school. She graduated in 2013 from the Health Care Assistant Program at the College of the Rockies in Cranbrook, British Columbia, Canada.

"Diana Clifton" was born in 1969 and lives in the United Kingdom. She is a creative writer and broadcaster, and her publications include erotic fiction contributions for an online magazine. As a young girl, she loved dressing up and still enjoys doing so today. While relaxing in the evening, she finds a glass of red wine is always welcome. She sometimes feels she spends too much time on the Internet.

Wendy Coates was born in 1956 and has been living in New York City for over 30 years. She is an active member of her church in Manhattan, loves theatre, and is currently an executive assistant at Citigroup.

Stefania Dimaio was born in 1989 and is an esthetician from Gloucester, Massachusetts. She loves reading and writing as well as learning new languages and spending time with her little brother. Though she enjoys her family, she does not hesitate to make new friends and expand her social life!

Jessica Esau was born in 1992 and graduated from Delta Secondary School in Ladner, British Columbia, in June 2010. She is now studying to be a medical office assistant at Douglas College and lives in Port Coquitlam, British Columbia. She enjoys singing and hanging out with her friends.

Camille Gaddis was born in 1991 and is from Wichita, Kansas. She attends Fort Hays State University in Hays, Kansas, and plans to graduate with a bachelor of science in nursing and a minor degree in Spanish. She is grateful to her family, friends, and Turner family for all of their support and adventures over the years.

"Rebecca Green" was born in 1981 and resides in North Carolina. She holds a PhD in biomedical science and is pursuing a career in biomedical research and communication. She is grateful for her husband and parents, who give her great strength and support.

Caroline Keras was born in 1986 and is a freelance writer from central Massachusetts. She graduated with a bachelor's degree from Ithaca College in 2008. She loves baking, rooting for Boston sports teams, and spending time with her friends, siblings, parents, and extended family.

Susan Lazar was born in 1960 and lives in northern New Jersey, where she loves spending her time with her husband, two great kids, and an adorable poodle. She received her undergraduate degree from Syracuse University and her MSW from the Wurzweiler School of Social Work at Yeshiva University. Susan maintains a private psychotherapy practice, and she is enjoying her life!

Jennifer Liu-Mormile was born in 1964, lives in Queens, New York, and is an accountant in an accounting firm in Manhattan. She still attends TS conferences and other social events in the Turner community and maintains her other close friendships as well.

Michelle Orofino was born in 1968 and lives in New Jersey. An educator, she graduated with a bachelor's degree in elementary education from the University of Montana in 1986. She is a voracious reader who also loves to write about her experiences—both involving and not involving TS! She loves cooking, yoga, and spending time with her husband, friends, and family.

Patricia Selway was born in 1960 and lives in Folkestone, England, where she is a health care assistant. She has been married for nearly 20 years and is grateful for her health and the ability to do what she loves best: caring for her family, working in the community, and serving in her church.

Caroline Skene was born in 1967 and lives and works in London. She spent her childhood in the inspiring countryside of East Lothian, Scotland, and after leaving Knox Academy attended Edinburgh University, where she obtained a degree in pharmacology. Several years of scientific research later, she retrained in IT, gaining a master's degree in computer science. She currently works for a global telecom firm during the week but indulges her creative side at the weekends for the Barbican, Europe's largest multi-arts and conference venue. Her guilty pleasure is watching *The Apprentice*.

Joy Webster was born in 1965 and resides in Quebec City, Canada, where she is a school librarian and ESL teacher. In her off hours, she writes, plays flute, and enjoys singing and reading.

Fiona Wilson was born in 1993 and wrote her essay while a senior at Westlake High School in Austin, Texas. There she enjoyed participating in Speech and Debate as well as the Student Council. She is studying psychology and child development at Barnard College in New York City and will graduate in 2016.

Editor's Biography

Claudette Beit-Aharon was born in 1956. She grew up in an academic family, which perhaps accounts for her permanent curiosity and enthusiasm for continued learning. She is from Newton, Massachusetts, where she lives with her husband of thirty-six years. Fluent in Hebrew, she taught English in Israel for five years. For many years she taught Hebrew in the United States. In recent years she has taught English as a second language to adults. She has three grown children: two sons, and a daughter who is one of the essayists. She became a grandmother in October 2012.

Essays

Xo

The Cat's Game

Fiona Wilson

Games end in one of three ways: you win, you lose, or you tie. Nowhere does a tie occur more frequently than in the simple pastime of tic-tac-toe. When it does, it's called a cat's game. For most, the X's and O's are just that—a game. For me, it's a matter of biology. You might say I'm a sort of genetic tic-tac-toe board. Each cell in my body has a single X chromosome, and where one would expect another X, I have an O; and the result is Turner syndrome.[2] Although this diagnosis by no means defines me, it has greatly impacted my personal development and perspective on life. Despite the medical and social challenges, I'm wiser, stronger and to some extent, happier, without one of my X-chromosomes.

As a kid I sensed that I was different—and didn't like it. My parents let me know something was missing, chromosomally speaking, from the beginning. The most obvious thing lacking was inches. My mom and dad presented growth hormone injections as an option. Without them, I was unlikely to grow taller than four feet five. But I resisted the idea of daily shots. They would represent yet another thing that made me different. I didn't want to be fixed. I wanted to be normal already.

Despite valiant efforts, I never convincingly projected normal. I saw things differently from my classmates. Most of the time, I preferred the company of adults, who generally seemed more accepting. This preference, consequently, earned me a reputation as a teacher's pet. However, most of my teachers didn't get me either. It didn't help that I could barely sit still in class, which left me feeling out of control.

Finally, my frustration reached a boiling point. Now, any experienced elementary school teacher knows the power of a reward. One of my teachers, Mr. Carlson, used licorice.

---

[2] Classic Turner syndrome was known as 45, XO. It has been reclassified as 45, X to end confusion over the use of O. See the glossary for further explanation.

Given my difficulty staying seated, of course none came my way. But I loved licorice. So one day, not being able to stand it any longer, I crept slowly in the direction of his desk. I hoisted the jar above my head and sang: "I got the power!" The class chanted "We want licorice! We want licorice!" I made a victory lap around the room, with Mr. Carlson not far behind. Hearing their cheers, I finally felt popular. But it didn't last.

On some level, I recognized that my behavioral problems stemmed from my relationship with Turner syndrome. Still, I resisted making the connection. And on top of it all, math proved incredibly difficult. I would eventually discover that there is something called Non-Verbal Learning Disability. It can include an impairment of spatial perception and often accompanies TS. This makes tasks like measuring angles, gauging distances, and drawing shapes particularly challenging. While studying geometric proofs, it came to me that I needed to use a similar process to approach math in general. In order to understand a problem, I had to break it down. By justifying my conclusion at each step, as I did with proofs, the whole problem eventually made sense. This understanding of my own learning style has ignited in me a passion for education and sparked my curiosity regarding the human mind. These interests have largely guided my subsequent work with children.

In time I put all of this in the context of my XO genetics. My childhood reluctance to accept my condition showed me the importance of confronting personal challenges directly. I've come to recognize that no such thing as normal truly exists. Everyone in some way feels out of control sometimes. Not until I acknowledged the differences that made me feel that way could I work with them and move forward.

After a while, I grew more curious about my diagnosis. Sitting at the dining room table one day, I nonchalantly asked, "Hey Mom, if I had a daughter, would she be an XO like me?" My mother replied, "Honey, you're not going to be able to have a baby on your own because of TS." All I could manage was a faint "Oh, OK." In my mind, the doll carriage I had frequently used to enact my motherhood fantasies had abruptly turned into a pumpkin. My curiosity had brought me back to square one. The lure of denial returned.

Still, I decided to attend several meetings of the Turner Syndrome Society. Yet some of the information put out by this group itself discouraged me. Some of the pamphlets suggested that TS women were less likely to learn to drive, graduate from college, or even get married. Nevertheless, I remained determined not to become a statistic.

A camp for TS girls at Pepperdine University provided the perfect opportunity to reinforce that determination. Bonding with this group felt seamless. The commonality of our medical condition played a part, but I believe something deeper was at work. A combination of shared struggles and a desire for support united us. The resulting relationships have helped me to embrace my condition.

Turner camp also proved crucial when it came to the issue of infertility. In one meeting with an endocrinologist, we discussed alternative means of having children, and adoption in particular. Each of us expressed confidence about our ability to love our adopted children. Collectively, however, we acknowledged that it would be a long road, and could grieve for our common loss together and move on.

My condition has offered unique learning opportunities that have shaped me as a person. For that I remain grateful. There have been gains. There have been losses. In short (no pun intended), it's been a cat's game.

The Unmentionable Condition

Joy Webster

In early December 2010, I felt inexplicably pushed to do some research on Turner syndrome after putting it out of my mind for many, many years. There was no particular reason to look into it then, but somehow I felt compelled to do it right away.

When I stumbled across a website looking for participants for a book about the experience of women with Turner, it felt like a sign that my story should be told. Writing this article has proved to be a cathartic experience, strengthening my relationship with my mother and giving me a new understanding of how TS has affected me.

At age six I was diagnosed with juvenile rheumatoid arthritis (JRA). A few years later, the doctors started thinking that there might be more to my JRA than was meeting the eye. Girls with Turner are susceptible to autoimmune disorders, and my doctors were suspicious. I was sent for tests starting at the age of ten or so and started seeing new doctors (geneticists), not just my rheumatologist. Nobody explained to me why. I was born in the mid-1960s when research on many of the implications of Turner syndrome was in a fledgling stage.

Growing up in an isolated northern community (population 2,000) in western Canada, any specialized medical treatments or tests required that my mother and I take time off from work or school, make a three-hour trip to the city, and stay in a hotel. The next day we would wait for at least an hour and a half at the rheumatologist's office for twenty-five minutes with him. I would dress again, then go to another part of the hospital, and do the same thing in genetics. Afterward, it was time to be seen by interns; then off to the lab for blood work, then to the pharmacy. It took all day, and we were stressed to the max when we left.

Each of the doctors would talk as if I were not there, or were an unfeeling slab of meat on the table. Undressing in front of each of these strangers as they talked to each other *about* me instead of *to* me, I understood only that my development wasn't normal. One group of interns looked at my hairline and laughed. The kids at school were already teasing me for being short and flat-chested. The doctors' insensitivity reinforced feelings of being defective and a freak.

I wanted to understand what was wrong with me. Nothing was explained in a way that I could understand. They didn't know how to talk to a girl in this situation, or didn't have time, or didn't take the time. When I was fourteen, the geneticist talked about how babies are made. The reason remains obscure. Maybe he sensed that no one at home had talked about sexuality and reproduction, and it was important that I find out. Yet in that conversation he never explained my infertility.

I eventually heard the name "Turner syndrome" when I was fifteen, but I had no idea what it was, or what the implications were for my health or my life in general. I needed to ask questions, but I was unable to articulate them. Why had this happened to me, and could anything be done? Would I lead a normal life or have a boyfriend? Nobody answered those questions in 1980.

In the following visit a few months later, I was told directly that I had Turner syndrome. The geneticist explained that I could never have children but he gave me pamphlets on high-risk pregnancies anyway. He pressed to start me on hormone replacement therapy (HRT), as I still had no breast development. My mother had resisted, or it would have been started sooner. She was very concerned that the hormones would have negative health effects later in life.

Beginning to look a little more like the other girls, even if I wasn't average height, made me feel better. We had to get my prescriptions filled in the city, because we couldn't let anyone in our hometown know! This made me feel like HRT was something shocking. It is good to know that most girls start treatment earlier, which must reduce the teasing that comes with such late-onset puberty. I dealt with these social side affects for many years.

The question for the doctor was "Since her ovaries don't work, is anything else missing?" Radioactive dye was injected to x-ray my internal organs—very uncomfortable. Nowadays, it's probably a simple ultrasound. There was some good news: my ureters, kidneys, and uterus were normal. But my ovaries were only small lumps of non-functioning tissue.

I didn't care about the good news. I wanted a future with marriage and children. Now this would never happen! I went into mourning and walked around in a daze for weeks. It was like a mid-life crisis at fifteen. My mother refused to talk to me about it. She wanted to push it away and not give it too much importance. My father and brother weren't told. The message I got from my mother's silence was that we shouldn't talk about TS, because it's shameful.

After a few weeks, my depression lifted. I wasn't ready to be a mother, and I knew it. At the right time I could adopt, and my future husband would certainly agree.

High school turned into college, which turned into my first job, which turned into further university studies. At thirty-two I moved to another province, found work, and started a new adult life far from my hometown.

There, I soon became friends with a wonderful man about my age. He had lost his wife to cancer. His children really loved me, and I adored them. He felt like the perfect match for me, and I would have loved him even if he hadn't had children. But after about six months, he stopped calling. Two weeks after his last call, I found out there was someone else (a woman who was a friend of mine) and that they were engaged!

Several people said that he was just looking for someone who was able to run after his kids. I cannot help thinking that arthritis, likely because of TS, cost me the life I could have had with him. As a girl, I was sure that because I was a wonderful person I would get married, probably right out of college. I did not imagine that I would still be single in my late forties. It took ten years to consider trying again with someone else. I have had other relationships, but I have learned that I need to be happy with or without a man in my life.

Some months ago, I was talking with my mom on the phone and out of the blue, she blurted out, "You know, I've never told your brother about the Turner syndrome. He doesn't need to know, and you shouldn't tell people about it." Yes, even all these years later, TS could affect my self-image negatively. Suddenly, I felt like an awkward pre-teen again, and my heart sank. However, I am now a middle-aged woman who is in the prime of her life. I face my very advanced arthritis with courage and dignity; I am successful in my career. I lead a life that other people tell me they find interesting!

I now realize the power that comes from being informed and speaking openly about TS. How could my mother not act ashamed, when I myself demonstrate the same attitude by not talking about having Turner? How could I care so little about my health that I learned nothing new about it for thirty years?

All I can do is change my attitude from here on in and keep up to date with the latest information. After ten minutes of research, I learned that ninety-nine percent of baby girls conceived with Turner are miscarried. I was astonished and humbled.

This knowledge makes me more determined than ever to contribute to society. I am committed to breaking the silence. I will no longer be ashamed to talk about Turner in appropriate contexts. I never want anyone to feel the way I did, or be embarrassed because a loved one has TS.

When I answered the ad for contributors to this book project, I was encouraged to come to New Jersey, where the Turner Syndrome Foundation was having its annual Christmas party. I drove for a total of twenty hours to spend twenty hours with other Turner women. It was more than worth the effort. I had never met anyone else with TS before. It blew me away to meet these women leading normal lives with husbands, adopted children, and *big* jobs.

This research and trip to New Jersey happened just before I went home for Christmas. I put my newfound commitment and confidence to the test and took advantage of the chance to talk to my mother face to face. The conversation brought healing to both of us.

I explained to her that being silent had led to feelings of shame. She explained that she was simply trying to protect my privacy. We continued to talk, and I discovered that she had felt guilty all these years because she thought that she had *given* me TS! I showed her an article describing how TS is not inherited, and that the reason that a baby girl's genes are anomalous is still unknown. She is still struggling to believe this, because for her it seems too good to be true. I reminded her about her non-responsibility a year later, and I think it helped.

When I struggle with my self-image, I remind myself of what I have learned, and how important it is to continue to learn. The fact that I am alive shows that there is a divine purpose for my life. There is much more to accomplish. I am walking this earth in spite of staggering odds against ever having been born.

View from the Yellow Brick Road

Wendy Coates

My name is Wendy Coates, and I am a fifty-four-year-old woman with Turner syndrome. Being small, wanting to earn my living as a performer, and making a life for myself in New York City all presented challenges. Some of them I've conquered; others I haven't yet.

I was diagnosed in 1969 at the age of twelve. I went for a physical prior to Girl Scout camp that summer when my family doctor noticed how underdeveloped I was. He suggested to my parents that I see a renowned endocrinologist at the Jefferson Hospital in Philadelphia. We lived in a suburb of Philadelphia at the time.

When I met Dr. Abraham Rakoff, he seemed an enormous, but very kindly presence. He was most impressed that my father, who was an aeronautical engineer, had meticulous graphs that he had done by hand, charting my height and weight along with that of my brother and sister. I still have mine to this day. After the diagnosis, my mom sat next to me on her bed and explained what it all meant. I think the most important thing she said was not to let anyone treat me as younger than I actually was. Now that I've heard other women's stories, I'm even more grateful for the love, care and grace my parents demonstrated during this time.

My family moved to Erie, Pennsylvania, the summer I turned fourteen. I was about to enter high school. As you might imagine, I was intimidated. In spite of that, I took advantage of an opportunity to join the ninth grade chorus. I also auditioned for the spring musical—*Oklahoma!* Well, not only did I get cast in the show, but landed the plum role of Ado Annie! All of a sudden I wasn't the new little kid who looks as if she belongs in grade school, but the really funny actress with the wonderful singing voice. During the curtain call of our last performance, when my classmate Bob (playing a marvelous Will Parker) and I came on stage to take our bows, the entire auditorium rose to its feet as we reached the edge of the stage. That moment changed my life. I now knew what I wanted to be when I "grew up."

Lots of performances followed in high school and college, and yes, I often played children. I became a theatre major and was accepted into the advanced acting program, which led to an apprenticeship at the Berkshire Theatre Festival in western Massachusetts during the summer of 1978. That was followed by a role in their production of *Carnival* the following summer. Finally, I made the big move to New York City on Labor Day, 1979. During this time, my body blossomed, thanks to what my mom called my magic pills. Although it was hard for me to accept the fact that I wouldn't be able to have children, I focused on the fact that there are a myriad of reasons why women do not have children. It isn't the stigma that it used to be, especially for women in the career I chose. My relationships with men had ups and downs, like those of any young woman, with more challenges coming from my profession than the fact that I have Turner syndrome.

The biggest show I was cast in was also the show that brought me a new awareness of TS, as well as many other forms of dwarfism. Did you know there are over one hundred and fifty different kinds? I spent a year and a half with the national tour of *The Wizard of Oz* playing—yes indeed—a Munchkin. The role was wonderful, with opportunities to be very funny, say Very Famous Lines and sing A LOT. The production starred Eartha Kitt as the Wicked Witch of the West and Mickey Rooney as the Wizard. We traveled all over the United States and Canada, bringing thousands of kids their first taste of live professional theatre. It was truly a dream job. The Munchkins were played by a mix of people with different conditions. Three other women and I were around the same height—four foot seven, or slightly taller. None of the others had TS. Four men and one woman were little people, so they were less than four feet tall.

Working and traveling with them was quite an education. The woman – Leslie – has a very rare form of dwarfism and is incredibly beautiful. The audience would literally gasp when she made her entrance in the show. We were talking during a break one afternoon and I mentioned that I had TS. Her response was something I've never gotten before or since.

"Oh, I thought so."

"Really?" I said.

"Yes. I have friends who have Turner and you remind me of them."

At that point in my life I had met only one other person—a young girl—with TS. Leslie explained that some women with Turner are members of the Little People of America Association and she knew them from the annual conventions that she attends. That conversation, and others that followed, changed my perspective on what it means to be smaller than average.

Grocery shopping trips with them were a revelation. I had never thought about going into a grocery store and not being able to reach *anything* on the shelves. It was also a marvel to watch what they had to go through to drive a car. My biggest complaint about being a small woman had been that I was not able to walk into a shoe store—even here in New York City—and find women's size four shoes. Leslie had no choice but to shop in the children's department. My difficulties were nothing compared to hers! But we also had many experiences in common, which was delightful and very moving for me. What they added to the show, as well as to my life, is immeasurable.

My friends, family, coworkers and most of all my Christian faith have worked to keep my perspective healthy. I am grateful and extremely blessed.

Hitting the Curve Ball

Caroline Keras

For most teenage girls, shopping is a joyous experience. For me, shopping could sometimes be a nightmare. I am very short but have average hips so finding pants that fit nicely can be an extreme challenge. I also have one fat foot, due to the fact that my lymph system does not drain properly. This makes it a trial to find shoes that fit my right foot and are the same size as the mate. When I succeed in discovering a pair of pants or shoes that are just right, I feel as if I have really accomplished something, rather than just gone on an ordinary shopping trip. Having Turner syndrome has presented me with many obstacles throughout my life, although some are more severe than others.

One percent of Turner syndrome fetuses survive to term, and one in approximately 2500 live female births has Turner. This knowledge makes fighting the obstacles I have to face different than it otherwise would be. It has made me open and excited about new experiences, strong in my convictions, resilient, and able to adapt to changing circumstances.

Being the oldest of four children, I had the privilege of being called the "big sister" for about eight years before my younger sister surpassed me in height. She is now twenty-one years old and five feet eight inches tall. I am twenty-three and an even five feet.

As a young child, the thought of being short bothered me. Why would my two younger brothers be projected to reach over six feet, (a milestone which they both reached), and I would have to fight to reach five feet with the help of modern medicine? Now, as an adult I know that I cannot escape my height, nor do I want to, but I have wondered, being on the taller spectrum for a Turner woman, how tall I actually would have been if I had been born with that second X chromosome. After many years and observations of tall people from a bench at the mall, or standing with friends at a party, I now realize that I would not be able to live up to my small and feisty label if I had two X chromosomes.

Growth hormone shots started at age four, though I was diagnosed with Turner syndrome at birth. I can remember sitting in a conference room at my endocrinologist's office, watching my parents learn how to give injections to an orange. I was told that the thickness of an orange skin is like the thickness of a human skin, so it is similar to the sensation of giving a shot to a person.

For the next eleven years, I took those shots faithfully every day though not without a few changes over the course of those years. I started off by rotating the injections between my arms and legs. That quickly stopped when the leg injections proved to be too uncomfortable.

Before a sleepover at a friend's house, my mom would give me a shot before I left home, instead of our usual routine of taking my medicine before bed. One of the proudest moments of my life was when I overcame my fears and gave myself a shot for the first time.

When I was twelve years old, I decided that I would like to learn more about the condition that I live with every day and meet other girls like me. I signed up to go to a camp for girls with Turner syndrome held at Harmel Ranch in Gunnison, Colorado. A little nervous, I boarded the plane from Boston to Denver that July morning by myself. It was my first plane trip ever. After arriving in Denver, I had to switch planes and take a puddle jumper to the tiny Gunnison airport, a total journey of almost two thousand miles.

From the minute I got off the plane in Gunnison I was surrounded by fantastic people and felt right at home. There was a week of eating, learning and laughing with about sixty other girls who had Turner syndrome. Having such a good experience at camp when I was a little hesitant in the beginning has made me eager to try other new things. I could now take some risks in hopes of experiencing life for what it is, an adventure and a blessing.

At the beginning of the week at camp, the nurse on site encouraged me to give myself my own shot. I promised to do so by the end of the week, despite how nervous I was. I knew that the nurse was just trying to be nice and that she would not even really be *that* disappointed if I failed to make it happen by the end of the week. But I am stubborn, so not fulfilling the promise was not an option.

Every day there was that walk to the nurse's office, with the anxious feeling in the pit of my stomach, as I said "not today". One more day had passed by to make good on the promise to the nurse and more important to myself.

As cliché as it sounds, on the last full day of camp, I pulled what arm flab I had on my little body over the edge of the table and injected myself with what I considered, at that time, my lifeline.

Exiting the door and running to the nearest pay phone (cell phones were not as available in 1998!), I screamed to my mom that I had done it! That little step toward independence ultimately helped me at other times when I did not feel confident enough to reach for what I wanted.

My next challenge was when I decided to go to a Catholic high school after having gone to public school with the same group of kids for eight years. I also started volunteering as a sixth grade religious education teacher. These choices have turned out to be two of the most rewarding experiences of my life. They gave me a chance to meet new people, impact their lives, and have them teach me a thing or two as well.

Having Turner syndrome has made me firm in my convictions and given me a strong sense of self. Being a little different from other children, I was often teased as a young kid. This gave me the opportunity to talk to my mom about what makes people act the way they do toward others. We discussed what good and bad qualities there are in people and, most of all, who I am and what makes me special. These early talks with my mother taught me that everyone is unique, so one has to embrace people for who they are. There are always going to be those who try to keep you down so you have to ignore them and be true to yourself.

These conversations helped me to establish my goal of how I want to be: a loving, caring, friendly, upbeat, helpful, and driven person. I also learned to associate with people who like me for who I am. I know how to get the facts, make a stand, and hold strong to what I believe.

Being five feet tall has sometimes proved to be challenging. I often have to get a stool or ask someone nearby to grab something out of a high cabinet. It is also important to keep my weight under control, because shorter people have trouble carrying extra weight well. Yes, I have to get my pants hemmed and drive a bit closer to the steering wheel, but there are definitely advantages too. I am lower to the ground for sports, which helps in maintaining possession of a ball or moving to get a ball. I will also be carded for at least the next ten years, and fit into children's clothes when the adult side is looking a little sparse. I can hide in smaller spots when I play hide and seek with my younger cousins or the children I baby-sit.

Turner syndrome has thrown me many curve balls that most girls do not have to try to hit, but it has taught me perseverance, flexibility, conviction, and what is important in life, including friends, family, and others who care about me. Learning all of this prepared me to face the challenge of college.

Passing the Torch

Michelle Orofino

When I sat down to write about how Turner syndrome has affected my life, I thought to myself, "How hard could this be?" Writing about TS would be nearly as easy as writing a shopping list. I have become so proficient at rattling off the most typical characteristics to the few people who have been curious enough to ask, or those who I have confided after a fair amount of caution and deliberation. The easy part was that I would be sharing my story in a safe, anonymous way, without the looks and questions that usually accompany an actual conversation. It might not be surprising to learn that it hasn't been nearly as effortless as I had thought. The project has helped me realize that TS goes beyond a list of common characteristics. My personal experiences amount to much more. This creative outlet has helped me to develop my unique voice.

This is not to say that I had never thought of myself as a little distinctive. From an early age, I refused to be confined to the typical little girl's pastimes of dolls and tea parties, if those things truly are typical pastimes. I'm not sure I knew that many girls who fit neatly into strict categories. Sure, I loved my Barbie Dream House as much as the next girl. My friends and I would spend many an afternoon making up adventures for our plastic playmates. But that wasn't quite as much fun as including our brothers in the mix and having a game of touch football, or racing our bikes around the neighborhood. Those pre-teen days were among the last times I considered myself equal to other children. I had no qualms about challenging a boy if I thought it was called for, and that even might include a punch in the nose. Of course, I usually regretted it later but that's a different story.

School could be a struggle. For some reason math didn't come as easily as reading and writing, and I ended up having to get after-school tutoring from my teacher. This not only cut into my free time but was a little embarrassing. None of my friends had to stay after school to get extra help. I also assumed that they were not up until ten at night, (sometimes in tears of frustration), struggling to get the required number of long-division problems done by morning.

Flash forward to the seventh grade, and math had not gotten any easier, nor school in general. My peers were growing taller, their bodies changing. I still looked pretty much the same as I had four years before, and my parents were starting to get a bit concerned. My mom and I had "the talk" and I assured her that even though my period has not started yet, it would happen very soon. I truly believed this. I had heard of late bloomer*s*, which must have been the reason why I had not sprouted up. However, at that time I began to tell myself that if I did not start to bloom soon, I wouldn't bloom at all.

I did not give up hope. Besides, being small had given me the advantage of remaining a little kid for a while longer. The adults around me seemed to treat me a little differently than they would other girls my age; a little younger, maybe not quite as independent and capable as an average thirteen year old. By this point, I had become fairly accustomed to this treatment. I just went along with things even if it was confusing and a bit humiliating.

Finally a routine visit to the family doctor turned out to be not quite so routine. He recommended that I see a specialist, a geneticist from a town an hour away. Soon after, I remember sitting in a large office, and being told that I have a condition called Turner syndrome. A nurse was there to draw blood for a test called a karyotype.[3] It was the only way to determine a firm diagnosis, but that doctor seemed more than sure of himself. Walking out of the office on that beautiful June day with summer vacation about to begin, I was sure life would never be the same. Lots of tears were shared. My mom and grandma were just as bewildered and shocked as I was.

---

[3] For more information please see the chapter on karyotyping.

We decided to make gallons of lemonade with this sizable lemon that was dumped in our laps. It occurred to me, that in spite of the fact that everyone was very supportive, that the brunt of this work would be up to me now. Boy, it was about the loneliest feeling in the world.

The brochure that the doctor gave me in the office was filled with as much information as was widely available in 1981, and provided a little comfort. Even though I would never bear children, hormone replacement therapy would help with puberty. It promised that there are many other women out there with the same condition, most of whom have normal relationships with men.

This pamphlet was illustrated with a drawing of a pair of legs, one pair much shorter than the other one, behind a tree with initials carved in a heart. It was cute to look at, but I was not so convinced that I could lead a fully normal life and maintain adult relationships. Besides, the relationships with my peers were already getting a little harder to manage, even before the diagnosis. I had felt more than a little alienated from my classmates since the beginning of middle school. In spite of this, I did keep a handful of friends through middle and high school. But I pretty much stopped participating in sports activities, something I had always enjoyed in grade school. Other girls seemed intimidating, and their skills and abilities far out of my reach.

In eighth grade, after a brief and frustrating stint in a mainstream algebra class, I got the extra one-on-one help that I needed. These special classes were affectionately known as "bonehead" math classes. It did not take me long to associate my own intelligence and abilities with the nickname. The fact that I did well in all other classes really didn't cross my mind at the time. A stigma is a stigma and a deviation from what's normal was not something I was ready to come to terms with. It became second nature for me to look in the mirror and see a living example of a diagnosis, a condition—at best a person who was seriously flawed. I had erroneously but thoroughly convinced myself that my options in life would be very limited.

My high school days were definitely a scary and confusing time. I managed to blend into the woodwork as much as I could while the effects of the hormone replacement therapy were gradually becoming noticeable. In many ways it was a relief to experience the usual rites of puberty, which were a reminder that my body is capable of doing the normal and typical.

Ultimately, hormones didn't help to ease the depression I sometimes felt, or to assure me a spot on the cheerleading squad, or to get me a hot date for the senior prom. In spite of my oh *so* sunny (not!) attitude, I made and kept a few new friends and even had a boyfriend by senior year. I'll never forget that first dance we attended together. He was so much taller than I, that when we had our picture taken that night, he had to kneel down so we could both be in the picture. While I was navigating the unsure waters of my first relationship, the idea of relating to him as any more than a friend was difficult.

Being so self-conscious of my physical difference just got in the way, even though he did not seem to be bothered by it. This was not exactly a winning recipe to maintain a relationship, so boyfriends were few and far between. How I yearned to meet that one special boy who would sweep me off my feet! It would take a few more years to meet him, so I had to learn to set my yearnings aside and get through high school. With the help and encouragement of family and a certain high school guidance counselor, I did apply to college and was accepted to a small teacher's college about an hour away from home.

It turns out college was exactly what I needed. The pressure of maintaining popularity and fitting in is essentially gone and is replaced by the task of managing hours of unstructured time without adults there to tell you what to do and how to do it. There are classes to get to and schoolwork to complete. Add to that the temptations of living in a dorm with dozens of other girls who have also just left home and want to celebrate their newfound independence with lots of partying, and you have got an authentic college experience!

Education became my chosen profession, not only because the thought of working with young children is appealing but also because I decided that the coursework with no calculus or physics classes required to complete a degree, might suit me better. At college, I almost immediately made some amazing friends. I discovered, through a wonderful English and drama professor, that theater is not only a worthy use of my time but that my height has absolutely no bearing on my participation and enjoyment. My introduction to the theater began with a position as a director's assistant for the first production of the school year. Almost at once, I fully enjoyed the responsibilities that came with the role. More than that, the more that I got to know my fellow theater geeks, I realized that none of them seemed to fit into a mold. They were not at all concerned with how other people perceived them. Their thinking was more on a "take me or leave me" level. It was and continues to be an empowering message. It has probably connected me with anyone who has ever identified themselves as an outsider or outcast, or even anyone who has never felt like just another face in the crowd. Let's be honest, who hasn't felt that way at times? It is much more interesting to be a distinctive individual. My theater friends expressed their individuality through their style of dress and through creative expression. I had the double advantage of not only dressing differently but also being physically distinctive.

Those college years were a time when I embraced my particular distinctions. In spite of my newfound wisdom, completely coming to terms with Turner syndrome was still a work in progress. At the same time, I continued to struggle and triumph in certain academic areas. A few professors were particularly encouraging about my writing skills, but I slacked off in other areas and had to work doubly hard to catch up. Back then, learning challenges were still a great source of self-doubt and frustration, although there were always supportive people there to help.

Today, I consider those struggles part of growing up and through that process, I gradually moved closer to becoming a full-fledged adult. I say "gradually" because it did take me a bit longer than most to reach that level of maturity and leave adolescence and all its concerns behind. Was it because I developed physically literally years after my peers did? Was it perhaps because of my own personal childhood experiences? We are all shaped to a large degree by those early years, and I am sure that my emotional maturity developed right along with my physical maturity. If that took a little longer than average, I have not come away from the experience any less worldly or wise than any other women my age. For a time, I had both the luxury and the disadvantage of peeking into the adult world from the outside. Even to this day, there are many elements of adulthood and maturity that I would just as soon leave to other adults. Which is not to say that I shirk responsibility or obligation; I have just come into my own version of adulthood.

More than twelve years ago, I met the man who would become my husband, after many years of wondering if there was anyone worthy of my time and my energies. Goodness knows, I had done my share of dating and searching. When I met him, I knew I had found the only man who would be up to the task of being with me, the whole package included. It has been a mostly sweet life so far. Sure the issue of children has been prominent, as it is with most married couples. Throughout my thirties the notion of not being able to give birth to a child did produce fresh waves of grief and sadness. Since an early age, infertility has always been something I had been stoic and matter-of-fact about. My husband and I discussed the various options of becoming parents, sometimes calmly and reasonably, sometimes not so much. As of today we have agreed to remain childless and it's something I am pretty content with.

How would I describe my adulthood? It is incredibly normal in many ways. Life can be a roller coaster, but I have been enjoying a fairly smooth ride. Fortunately, my health has been good. I am adamant about staying physically fit and keeping up my HRT regimen.

If I could include one thing here that has been the greatest benefit of being a woman with Turner syndrome, it would be meeting women who are similarly affected and becoming an active participant in the local chapters of the TS Society. What could be more inspiring than meeting other women who have taken their diagnosis and run with it, not looking back, nor letting TS hold them back? This has probably done more than any time in a therapist's chair could do. It has helped me come to terms with Turner syndrome. I not only accept it, I can celebrate my personal condition. I embrace it as a part of me, just not the whole me.

In my forties, to my delight, my eyes are being opened to a wider horizon, much broader than the one I had previously seen for myself. My strengths are becoming more focused, and I am able to use them more than ever, both professionally and personally. To those young children and teens just faced with a diagnosis and parents who may be overwhelmed and feel powerless, I would like to be one who passes that torch. All they need do is run with it.

Expect the Unexpected

Camille Gaddis

Is it possible to appear smarter than you really are solely based on your size? In my case the answer would be yes! I was treated like a normal child who happened to be smaller and cuter than the rest. My mom recalls how some people reacted to me when I was an infant and we were at the grocery store. I was a perky nine month old sitting in the shopping cart, but to everyone else I looked like a six month old. "Oh, she is so smart! Look at her sitting up. She is a genius, just look at her." When I was a toddler, my parents kept me active and independent with music education, reading activities, and group play. There were times when I would be talking to an adult who would say, "My, you sure have a grown-up voice. You say a lot of big words too. Are you really as old as you say you are?" The maturity of my voice and vocabulary were exactly the same as other children my age, but I appeared extremely advanced. It was so unexpected that someone in such a small body would have my intellect. These experiences helped build my confidence. My parents always encouraged problem solving and socializing, which helped me to be successful as a young adult in nursing school.

My mother's pregnancy was normal and without suspicion of any health concerns. Though not sure of the date of conception, based on my development on the sonogram, the doctors estimated that I would arrive on May 4th. But I was a full-term baby when I was born on April 23rd, weighing in at five pounds, six ounces. The only concern was my puffy left foot. The in-house pediatrician ordered an x-ray and examined it. He determined that there was no deformity, just some swelling. Perhaps I had lain on the foot at an unusual angle in the womb? This was probably the cause. He concluded that it would go away in a few weeks.

Mom and Dad brought home their bundle of joy and made an appointment with a recommended pediatrician. The infamous puffy foot that had become an accepted part of me by my parents was a red flag for our pediatrician. He ordered a karyotype, bringing on emotions of terror and confusion in my parents. The possibility of problems with their little girl was just too much for them to bear. After the test, they wrapped me up and took me home and wondered what new hurdles they would face when the results came back. They had entered a personal world of genetics that had never before crossed their minds.

The news came in a phone call from the pediatrician, who simply said, "Your daughter has Turner syndrome, a disorder that affects females. She will be small in stature and infertile." My mom went to the library to find out what she could about this mysterious disorder. All she found amongst piles of medical textbooks were two paragraphs about a 1950s case study of women in the South Pacific, who were short in stature and of diminished intellect. After reading the little information available several more times and finding no hope or comfort, my mom laid down her head and cried about what she thought my future might be like.

Then my parents visited a genetics specialist at the Kansas University School of Medicine-Wichita. He shared insights and information on recent findings, testing, growth hormone therapy treatments, and other physical things to expect. All of it was medical. There was no awareness or suggestions of support groups or counseling on the social aspects of raising a girl who has Turner syndrome. Life went on, and my parents cared for me as they would any other child. But they were still grieving, and they had trouble accepting my diagnosis out of Fear of the Unknown.

My mother says that though she loved and accepted me unconditionally, she struggled with the diagnosis until Christmas Day that year, when she read an article in the local paper. It highlighted a young teen from Kansas who had been awarded the Horatio Alger Scholarship for overcoming obstacles. She had Turner syndrome. She was also an honor student and a cheerleader and had plans to attend college. Her outlook on the condition was very positive. "I have Turner syndrome. I'm like missing a chromosome." After reading the article, my mom's grief lifted. She was comforted in knowing that hope existed, and that as a family we could make it through any obstacle.

Another important aspect of my upbringing was the diversity found in my preschool playmates. They were each unique in their own way. We were all different sizes, colors, economic backgrounds, and cultural upbringing—all playing in harmony. So, other than having a few more doctor's appointments than my friends, and having no sibling to be compared to, I never thought of myself as unusual, although there were, and continue to be, instances when my size is both helpful and surprising.

I remember as a kindergartner that the door to the supply closet was locked, but there was no doorknob; my little hand was the only one in the class that could fit through the hole and unlatch the lock.

Then, at age six, a whole new chapter of Turner syndrome opened up for me: growth hormone therapy. I think that my parents had more trouble with the experience than I did, but they managed to play down their anxieties. They were strong for me despite their own fear of needles. They knew how important growth was for my appearance, confidence, and how I would relate to people. I have a distinct memory of a nurse coming to the house to show my parents how to give shots. My dad used a banana to practice. I laugh about it now, because by the end of the day the banana had turned completely black. I look back at that time and think how much courage my parents had. They controlled their own fears and met this challenge head on.

As the years went on, the height gap between my peers and myself was still substantial and increased little by little as we got older, but it would have been much greater if I had not started growth hormone when I did. At age twenty-one, I am a proud five feet, not much shorter than most of my relatives, or many of my peers.

Since most of my family members are short, I would have been short without Turner, but every bit helps! I would also like to encourage girls to learn to give themselves their own shots. You have no idea how liberating it can be. Most of the time it is also less painful than when someone else does it. It isn't always perfectly painless of course, and there are the occasional slip-ups. I remember one night when I was getting the syringe ready. I had a hand spasm, the syringe flew out of my hand, and the needle landed on my big toe. Ouch! Things like that can happen, but you do not let fear take over and get in the way of what needs to be done.

When I was fourteen years old and planning a trip to Indianapolis, Indiana, to spend a week with relatives (without my parents), I was prompted to learn how to give myself the shots. It gave me a whole new sense of freedom and a feeling that I could do anything.

However, it was really difficult not to go through puberty at the same time as my friends and peers. There were times when there would be conversations about periods and puberty with girls who didn't know about my condition. I felt the need to lie to fit in. It really crushed me, because I felt like a fraud. Whenever a girl complained about cramps, I would agree and say, "I know how you feel, girl; cramps are the worst!" Is it possible to be at all convincing without personal experience of puberty? It seemed to work, though it didn't feel quite right. I gave information based on what I had heard. Fortunately, my parents, family, and close friends were supportive and kept me comfortable until I finally began taking estrogen. Undergoing hormone replacement was a very important experience. It helped me become a woman and feel closer to my peers. It was good to develop breasts and other mature womanly characteristics, even though the changes were made artificially.

A few years later, in college, I started taking contraceptives to begin menstruating. After the first cycle, I felt relief and inner peace, finally understanding what the other girls had been talking about all of those years. I finally felt more complete as a woman and was comfortable in my own skin.

One thing I regret is not being consistent with my growth hormone therapy in the last two years I took it. Sometimes I wonder how much difference it would have made in my height. When you have given yourself shots of growth hormone for eleven years, as I did, it begins to get a little tiresome, and you start to find reasons for not doing it anymore. Nonetheless, I encourage all growth hormone veterans and girls just starting out to *stick* with it until your doctor says you are done growing. You never know your true potential until you are finished with treatment. You might even exceed your own expectations, especially if your close relatives are tall.

In middle school I kept busy by playing an instrument in the school band and participating in student government. In high school I continued with band. The school year was full of performing at Friday night football games, marching in parades, and attending musical festivals around the state. I also wrote for the school newspaper and became an editor during my senior year.

There are times when you have to hold your head up high and demand some respect, but I take the time to laugh at myself and what makes me unique. When I was at a restaurant recently, the hostess gave me the kid's menu. Then the waitress gave me a kid's soft drink! But I calmly ordered beef enchiladas from the adult menu and then colored the pictures on the kid's menu with the crayons that I was given.

Many students were tall, but I have never let this faze me. When I was honored at a high school event, a group of teens was called to the stage. I ended up in between two very tall girls. They looked at me as if to say, "What are you doing here?" I smiled at them and thought, "because I belong here." I did not even consider moving. My place *was* center stage!

Whenever I had a doctor's appointment during school hours, my dad would be the one to take me, because he worked the evening shift. I really enjoyed having that special time with him. We had a tradition of always going out to Wendy's afterwards. Sadly, my dad died in a traffic accident when I was in high school. But I treasure all the many happy memories I have of him and all the good times we spent together.

In college I don't stick out anymore because of my height. It is a medium-sized university, and there are so many students that everyone is different. I am very busy with studies, but when I can, I enjoy attending school activities and sports events. Another great thing about college is that it fosters an environment that encourages individuality and pride in yourself. I joined the Hispanic American Leadership Organization because of my Hispanic heritage. But in general, it is important to be proud of who you are and enjoy the things that make you special. Do not let your height or other Turner characteristics bother you, because most of the time other people don't notice.

Life can be hard even when you have people who are there for you and love you unconditionally, but I cannot imagine going without the support of a TS community. It is wonderful to be involved with support groups, community organizations, and (if you are between the ages of thirteen and nineteen) the annual Turner summer camp. Even if you don't think you need that much support, you can help others and help yourself at the same time. There are thousands of other families in need of financial assistance, medical information, or just a kind word. I know that I am not alone; I have met amazing people who have changed my life for the better, whom I would have never met if I didn't have Turner.

Sometimes it is difficult for me to accept that I will be unable to bear children, but I can still become a mother through the miracle of adoption. Although the path is quite different, I will end up in the same joyful place that most women do, when I am older and the time is right. I pray that other girls and women who will be unable to get pregnant find the same peace that I have found. And I wish those who *are* able to all the luck in the world.

I know that if you let disabilities and challenges get in your way, they will undoubtedly do so. But if you go through life doing your best, nothing can stop you. Even though there are bumps in the road, with friends, family, and your Turner family for support, you can overcome anything. Your doctors and nurses are also a part of that support group. They are people who know Turner inside and out and will answer all of your questions, as well as steer you in the right direction. Take life one doctor's appointment at a time. Always remember that you are special. We are on this earth for a reason, and we should never forget that.

Let's Talk about Sex!

Diana Clifton

The subject of sex in the TS community and at conferences is almost taboo, and this attitude toward something so central in life is perplexing. It is time to be bold and delve a bit deeper into the subject. I think we need to talk about sex more openly and see the what the conversation reveals about how we feel about ourselves as women. Thinking about my personal journey to a positive relationship in my forties, enjoying sex, and having Turner syndrome, I hope that what I share will prompt discussion and debate. Not everyone reading this will agree with me, or have similar experiences, but let's talk. Of course, each person has to find what works for her.

I was first asked out when I was almost eighteen. I was so excited! He was an intelligent, very good-looking boy, and we went for a walk by the river. I was shy and unsure of how to act, and don't think we even kissed. After that, we only spoke briefly at school, both of us glancing down at the floor, awkwardly shuffling our feet.

Having been asked on a date did give me more confidence, and not long afterwards at a party, I met a friend of a friend's brother from another town. We talked and flirted for what seemed a long time. After a while we snuck out to the garage where I sat straddled across his thighs kissing him deeply. I enjoyed the reaction, getting him worked up and his breathing quickening. He made a very genuine, natural comment that I had a nice body, which was a pleasant boost to my ego. We kept getting interrupted, so we took two waterproof coats that were hanging up on the wall and went to find somewhere quiet. I remember clearly thinking 'I might have sex for the first time tonight!' It was a very conscious decision. Choosing someone I might not see again ensured that there would be no embarrassment afterwards if things didn't go well—and of course there was a large element of curiosity.

Despite a lack of experience, I was feeling pretty confident and wanted to explore. Whilst having no specific expectations that night, I also realized that there was a good chance that it would be less than ideal. Anyway, we walked up a narrow grassy lane in the darkness and after a few unsuccessful thrusts I returned to the party, extremely muddy and still a virgin. Being inexperienced and not knowing what to do, I still somehow understood that the excitement of the moments in the garage had passed. I didn't see him again.

I tried again later that year with someone I met regularly on nights out with friends, and he did manage to enter me a little way. But the circumstances were not conducive to being relaxed or speaking openly, and I simply chalked this up to experience. After those first unsuccessful attempts, I wondered whether my vagina was on the small side or perhaps shaped slightly differently than normal, as I had been examined only externally at diagnosis. Having only felt a penis with my hands, I found it huge. I wondered how on earth that was going to work. Thankfully, I dismissed those concerns. I wanted to find and nurture a solid and genuine relationship and saw sexual competence as an integral part of being close to someone.

Full intercourse didn't happen until I was nearly twenty, with my first serious boyfriend. We had been for a walk at a local beach during the day and spent the evening in his flat. He very much knew what to do, and what he wanted, which of course I did not. I remember him using his fingers inside me. He was very comfortable talking about sex without embarrassment. However, it was very much that he had sex with me, rather than us exploring what we both wanted. Although there were times when sex with him felt very nice, I never had an orgasm and never took any initiative. Years later I understood that I had been incredibly passive and did not assert myself at all in that relationship. My parents and friends never approved of him, and I broke it off after a few months. There were many tears, but I had learned to be less self-conscious, to recognise when I was allowing someone to take advantage of me, and to stop that from happening again.

The worst sexual experiences were those times lying beside someone when I knew it wasn't working anymore and the emotional connection was gone, and thought "I just want this to be over."

The best experiences are when there is a really strong bond. Since we are able to really talk and feel at ease together, the relationship is what enables great sex. Then, in the moment, nothing else matters, and both partners are completely focused on each other.

For me sex goes beyond the ordinary when I am adventurous and push my boundaries. I try new things that feel natural but exciting and that I am comfortable with, like dressing up in an outfit that makes me feel sexy and beautiful. It could be a glamorous dress or special lingerie. I quite like my legs and wear something to emphasise them. Sometimes I feel sexy if I dress down in a T-shirt and slinky knickers.

After being good friends with a man for a couple of years, a romantic relationship evolved. Sex began slowly, and we spent our initial months as a couple occasionally sleeping together but only kissing and touching. It felt unrushed, and during one of these evenings I had my first orgasm. The sensations took me completely by surprise but were wonderful—like a gentle vibration through my whole body. I immediately told him it was the first time I had experienced this. Over time I have learned how my body responds and exactly what touch and associated thoughts bring me to orgasm. It happened because I was incredibly relaxed, and I wasn't thinking or worrying about having an orgasm (or not).

I have on occasion, suffered from cystitis, which is a common bladder inflammation. In my case it was definitely as a result of irritation and bruising of the urinary tract during especially enthusiastic sex. I would wait anxiously during the following twenty-four to forty-eight hours for that burning sensation and pain. Sitting on the loo at three in the morning squeezing out a few drops of hot, stinging urine was awful. I read up about cystitis. General female anatomy means women as a whole are susceptible to both cystitis and urinary tract infections.

Knowing how to deal with a bladder infection is important for any woman, but particularly if you have horseshoe shaped kidneys. At the slightest twinge, its time to start drinking lots of water. Cranberry juice can also be good. I have also tried stirring sodium bicarbonate into water which unfortunately is a disgusting drink. The important thing is to flush out bacteria quickly so you will be less likely to need antibiotics. It also helps to empty the bladder after sex.

Once in a while I have found certain positions unpleasant. I say something and move to feel more comfortable. Sometimes something funny happens—so I just laugh! Sex isn't always perfect. Understanding and appreciating this as a couple, we treasure the times when it is perfect, but good can be quite wonderful too. Every relationship is different, but with a high level of trust comes the confidence to suggest things without concern.

Many women have issues with their weight, and I am not immune to those either. I eat healthy foods, generally do not snack, and exercise when possible. But if I give it a miss for a little while, I try not to beat myself up about it. Being comfortable with your body is not about what the scales say or a dress size. A sexual partner is often most turned on by the enthusiasm and interest of the other person.

There was a time when I had major issues with my breasts. They developed to different sizes, and at twenty-two I had a small implant inserted on one side to even them up. However, ten or so years later I had large sagging breasts heading steadily toward my waist which were totally out of proportion with the rest of my body. I opted for a breast reduction, which enhanced my confidence. I now have some scars and differently sized areolae but I don't care.

When I was about twenty-seven I found out that I have a section of Y chromosome in some of my cells. That somehow made sense to me, but I don't know if my attitude to sexuality and initial approach to sex would have been different if I had known earlier. I don't believe so, and certainly hope not. The fact that I wasn't concerned when I did find out makes me think it would not have made a difference. Infertility doesn't make me feel less a woman.

Sexually, I am interested in men. I have given some thought as to whether I also find women attractive, but it is only curiosity. I know women with Turner syndrome who are lesbians. I have never asked them how or when they came to understand this, or when they felt they could be open about it. The lesbians I know are very self-aware and self-confident. In my opinion comfort around the issues of sexuality is very important

When I was younger, I was searching for a good relationship yet didn't see marriage and children as my goal. Did that make me accept mediocre relationships on a couple of occasions? Probably. With mistakes you learn—which I did.

I also do not think I consciously thought "Oh I can't get pregnant. Great! I can have as much sex as I want!" Not long after I became sexually active, HIV was discovered, and the importance of using condoms was clear. There were awkward conversations around ensuring protection, but no decent person would object to condoms.

I have not told every sexual partner about Turner in detail, but I have explained that I have a medical condition, which is why I am shorter than average and cannot have children. It is not necessary to explain further.

What is the impact of HRT? Does it contribute to how I feel sexually? It must contribute in certain physiological ways. But I think there's a base level of libido and on top of that an emotional response. Hormones may not be at their highest, but the situation can be intense enough to supply the interest.

Here is what I would say that I learned over the years:

If a woman has any concerns about her genitalia or the health of her internal sex organs, an examination by a gynaecologist is extremely helpful. It is even harder to take the emotional risks that come with an intimate relationship if there are concerns that sex might be dangerous or physically impossible.

I let my body do what it does naturally, and if I am not aroused, well, there may be a number of reasons for that. Sometimes I may want to add extra lubrication. I have used some lubricants and been impressed. They can feel amazing and nudge things in the right direction. I would recommend the more natural options though rather than the 'tingle' ones if trying these for the first time.

When nervous or apprehensive, muscles tighten up and sex becomes awkward, and the uneasiness tenses the body still further. With sexual arousal, the vagina expands to accommodate pretty much any size of penis, but most certainly some positions are more comfortable than others. The only way to find out is to try, and, if its not comfortable, to have the confidence to say so. Anyone worth building intimacy with will want to know.

It is important not to be embarrassed about talking about what does and does not work. If we are happy lying together and entwining legs and arms, then that's wonderful. If I want to be more proactive about moving around, then that is fine too.

Sometimes there are difficult but important conversations, but they are less scary, and more successful, if initiated outside the bedroom. Both physical and emotional intimacy is all about good communication. I say what I like and do not like, and I am clear about my boundaries. I am not afraid to explore but not afraid to say *no*.

I try not to have specific expectations of sex or myself, or to worry about not being interested in sex all the time. The sex drive varies, and intimacy can be built up in many different ways within a loving relationship. Often when I was younger, I had concerns, and wished I could have shared them with a trusted friend, or spoken to a medical professional. For reassurance I read books.

After a decade long relationship, as well as several years of more casual dating, because of all I have learned along the way, my current relationship is extremely special.

My goal in sharing these thoughts is to encourage openness about issues of sexuality and sex. Women with TS should feel as free as other women to ask questions and discuss the issues that impact this important part of life.

Fragments Of Me: A Photo Essay

Brigitt Angst

Xo

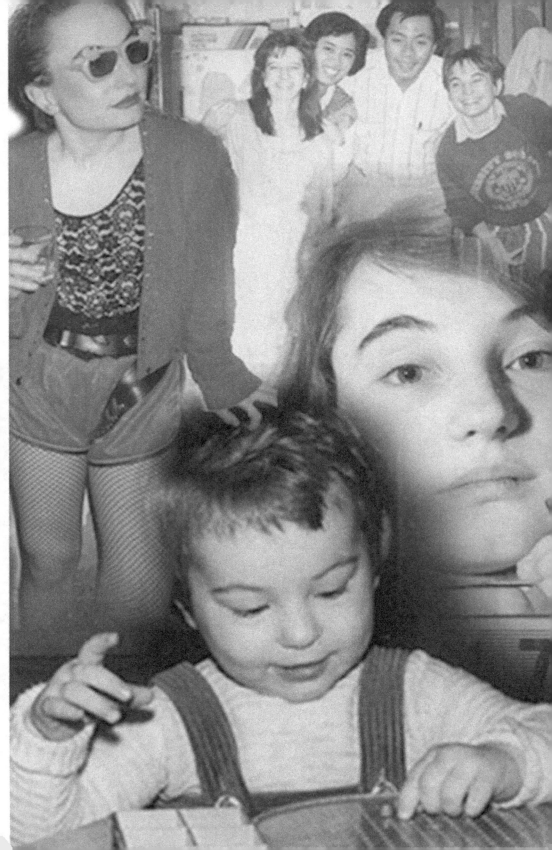

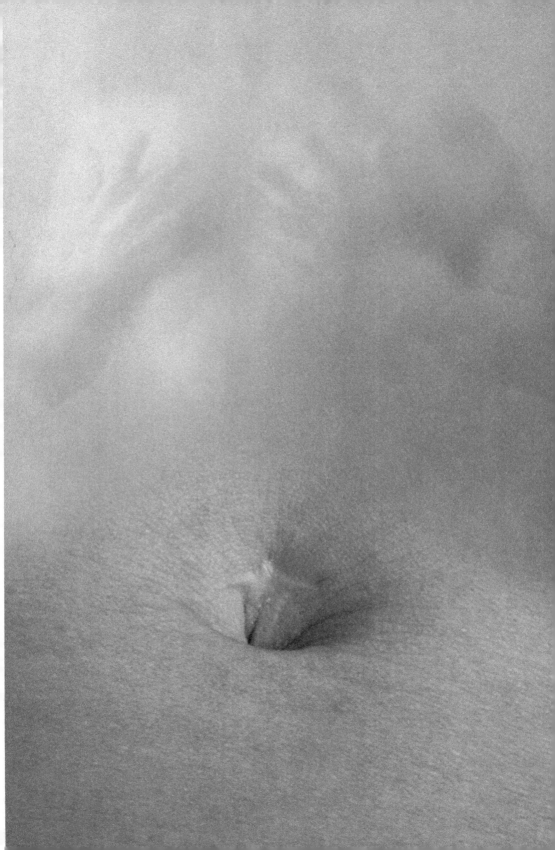

You can see the whimsical, inquisitive girl who enjoyed solving puzzles and dreamed of becoming a vet and healing animals. I loved pretending I was a boy, and preferred boy's clothes for years. Boys seemed to have much more fun.

Here is the brooding high school girl, looking so much younger than her peers, wondering where she belonged, stifled and isolated at an all-girls' school.

There is the postdoctoral researcher, with the best boss ever and amazing lab friends.

Many signs of TS are hidden, but in one photograph my elbow angles away from the midline of my body (cubitus valgus), a very common sign of TS.

See the laparoscopy scar in my navel? It is a reminder of the medical investigation carried out to assess the condition of my non-existing ovaries.

Who am I, and where do I fit in?

I grew up in a doctor's family in Lucerne, Switzerland. At the age of fifteen I seized the opportunity to apply to Atlantic College in Wales and at sixteen left my all girls' school for a high school with three hundred coed students from fifty countries. This transported me to a new universe of possibilities. The diagnosis at twenty explained what I felt inside. The subsequent hormone treatment, was a rocky road - becoming a fully developed woman at an inappropriate age.

I never wanted to become a mother and have always focused on other plans. The dream of becoming a vet transformed into studying biology. At twenty-nine with a PhD in Clinical Immunology, I left my native Switzerland for postdoctoral work in Chicago and London.

As the passion for science dwindled, in 2001 I decided to focus on my lifelong interest in photography. I completed my Masters in photography in 2011 with multiple projects on TS.

Whilst I experienced my first French kiss at sweet sixteen on a trip to Poland visiting a school friend, I had no interest in exploring sex. After my diagnosis four years later, taking female hormones awakened sexual desire. A one-night stand broke the ice and opened the door to happy and sexually fulfilling long-term relationships.

Plucking up the courage to change paths is never easy or straightforward. My journey has been fascinating and very rewarding: from science to photography, yoga practice, and a wellness business.

All humans are mutants somewhere in their DNA, but missing a large segment, or an entire X-chromosome, is far from the norm. Girls and women with TS are often encouraged to follow traditional feminine roles, such as work in the caring professions, or to aspire to a traditional family with husband and children. The media present us with an often exaggerated pink and frilly outlook on femininity, which I never identified with. This push towards "normalization" can cause a lot of tension and stress.

There is no denying that I could not have my own genetic children, and though I was not interested in doing so, this fact still has implications. We can have families in different ways, but we are removed from the natural life cycle.

Needing artificial hormones to induce sexual development was embarrassing. So I stuck my head in the sand. I got on with my life. I put the TS genie firmly in it's bottle and covered up this invisible lack of ovaries and fertility. Only in retrospect did I realize how my silence alienated me from myself for such a long time. In fact when asked recently what the worst thing was about having TS I replied without hesitation: "Not being able to talk about it."

I lived an outwardly successful life as a research fellow, and never felt captive to my DNA, but I was living in my head, disconnected from my feelings and sources of true inner happiness. Owning up, coming out, speaking out, facing and accepting TS is what has ultimately set me free and made me whole.

Embracing Scientific Inspiration and Faith through Turner syndrome

Rebecca Green

For the most part, my life has been pretty normal: prom, college, graduation, and marriage. It has been filled with both successes and challenges. But reflecting on the last twelve years in more depth, my diagnosis and coping with Turner syndrome has accelerated my emotional development and maturation, and given me purpose, conviction, and inspiration. I feel strongly that this personal growth and strength would not have been possible without the support of my amazing parents and my wonderful husband, whom I adore with everything I have.

I am humbled and awestruck each day by the intricate complexities of biology and in all aspects of God's beautifully designed creations. While science may describe genetic disorders such as Turner syndrome as mistakes in DNA replication, I have learned to appreciate the grander design, as well as the beauty and strength, that come with challenges, coping, and hard work.

My diagnosis inspires and challenges me in countless ways, every single day. My initial diagnosis generated the curiosity to understand the complexities of human biology through research of my own. I am pursuing a career in science, specifically Neuro-endocrinology and Biochemistry. Disciplining myself and working very hard enabled me to graduate cum laude with a bachelor's degree in biochemistry and I received my PhD in biomedical science with a concentration in neuroscience. I hope to help other girls and women cope with a TS diagnosis, and encourage appreciation for the beauty of biology seen in all details of God's astonishing and varied creations. Everyone, including those with Turner syndrome, should fully and passionately embrace their unique genetic combination and strengths.

I did most of my growing up with no knowledge of Turner syndrome. Except for short stature, there was no physical indication to prompt suspicion until my reproductive development was delayed. I was officially diagnosed at sixteen. I remember collapsing in a puddle of tears on my bedroom floor on the afternoon that the nurse came to our home to show me how to inject myself with growth hormone. This was during the week of my sweet sixteenth birthday. Self-injection seemed a completely insurmountable challenge, like a segment from the show *Fear Factor*. I will never forget the concrete strength and foundation that my parents provided throughout my first years of treatment and coping.

In that early and difficult adjustment period, my daily syringe of growth hormone was laid out on a napkin with a handwritten inspirational quote, notes of love, and silly stickers to make me smile and give me strength. I am filled to overflowing with immense gratitude and admiration for my amazing parents. They rewarded my persistence and courage by commemorating and celebrating each year of progress on the anniversary of the day of my first injection. That first year they gave me a bracelet, which I wore very proudly. Those tokens meant the world to me.

The shots, almost exclusively alternating in my thighs, got easier as time went by, becoming more tedious than anything else. I continued with them even as a freshman in college. With little privacy in a college dormitory it was difficult at first, but my family and newfound friends gave me comfort. My parents even continued to pack the supplies with encouraging napkins and stickers that entire first year. My roommate also asked me for warning so that she could leave the room when it was time for the shot, which was good for both of us.

My bone age was monitored regularly (with x-rays of my left hand) throughout the first years. I was lucky to still have room for growth in my bones as a teenager, having been diagnosed relatively late. At five feet tall, I look like any other petite woman.

I did have eight periods spontaneously between the ages of thirteen and sixteen. This erratic and delayed onset of puberty is what initially prompted my consultation with an endocrinologist. Hypothyroidism did not explain all of the endocrine and growth issues that I was experiencing, so my endocrinologist ordered a karyotype. Sure enough, four out of thirty of my cells examined were missing a piece of my second X chromosome. My final height was somewhat enhanced with genetic contributions from both of my very tall parents.

I remember getting very upset when x-rays were taken (bone-age or dental). There would be probing questions about the chance that I could be pregnant. Some technicians were not very considerate and would ask simple but awkward questions, like, "When was your last period?" They would poorly handle my answers of "over a year ago/six months ago" showing little understanding or sympathy. Technicians should be better trained and more tactful. They need to understand that many girls and women know that pregnancy is not the reason that they are not getting their period.

After four years of growth hormone between the ages of sixteen and nineteen, I had added three inches to my height. To celebrate and commemorate that milestone, my mother had a notebook printed of all the saved quotes from the napkins for each injection. It is one of my most treasured possessions.

Here is a sample from His Holiness the 14th Dalai Lama: "Never give up; no matter what is happening; no matter what is going on around you; never give up."

My transition to estrogen replacement therapy was a relief. I was finally finished with the injections. Hormone replacement clicked everything into place. I felt that my body knew that it was supposed to cycle, and with that change life seemed much more normal. Taking a few pills with breakfast every morning seemed easy when compared to a puncture with a needle.

Hitting the fast forward button, I graduated from college. While pursuing my PhD in biomedical science I took the plunge and tried an online dating service, and found a wonderful man very early in the process. We continue to share a great romance together, from camping under the stars, to watching the sunset from the top of the Empire State Building in New York City, to honeymoons in the Caribbean and Hawaii. We said "I love you" for the first time amidst the fireworks on the Fourth of July. He proposed to me in a beautiful garden adorned in snow and white Christmas lights, saying, "I will never let you go." After a magical year of dating and nine months of engagement, we were married on a lawn overlooking Narragansett Bay in Newport, Rhode Island where we had shared our first weekend escape together. On our wedding day, we both fell head over heels in love all over again and cried with happiness. My experience has been that true love soars much higher than simple romance embracing imperfections and treasuring the beauty and strength within. There is no greater comfort than such companionship; I now understand my grandmother when she speaks fondly of the art of handholding after more than fifty-five years of marriage.

That brings me to the next and most significant stage of challenges and coping with Turner syndrome. It was extremely gut wrenching to share my diagnosis and infertility. While we were dating, he knew that I had to take prescriptions for various things daily, for my high blood pressure and an under-active thyroid gland, etcetera. I jokingly say that I am already a little old lady taking lots of pills. But while we were dating, I did not tell him everything about my diagnosis at once, but incrementally in small pieces. You have to carefully assess a boyfriend's response to prescriptions and appointments with specialty doctors, etc. Mr. Right will be understanding and kind.

Soon after we got engaged, I finally amassed the courage to tell him that I will not be able to get pregnant without medical intervention, if at all. I vividly remember that initial conversation about my diagnosis and planning to start a family, which happened on Super Bowl Sunday. Thank goodness he is not a football fan. His response could not have been more comforting. He was happy and relieved that there could not be an unplanned, poorly timed pregnancy. His response was, that there are so many children in need of a happy home. "Why would we risk challenging your body to get pregnant in the future?" To this day, my husband has a wonderfully supportive and immensely protective attitude.

At this stage in life, I want to start my own family and want to raise a child with the love of my life. My heart aches when I see friends of mine get pregnant and start their families with ease, or when I see pregnant women shopping or young families walking with a baby carriage. But I share my life with a wonderful man, enjoy every second with him and am successfully pursuing my scientific and academic work. I hope and pray that within a few years, (after our thirtieth birthdays) we will rescue a baby in need of a home and become happy adoptive parents. The movie *Juno,* depicts many details of my dream of adoption and touched my heart in a very special way.

Somewhat small by comparison, is the other challenge that I face in increased dental bone loss. This is because of a childhood accident. I shattered my two front teeth at age seven. One of my front teeth was replaced with a dental implant while I was in college. The other front tooth has been a work in progress for almost two years, as it became badly infected as I was finishing my graduate studies. As the result of the accident, I have had five root canals (two had to be redone) and six oral surgeries with bone grafting etc. So I have had much more than my fair share of dental surgery before the age of thirty. The process has been not only painful, but also arduous and expensive. It is still not over. A search of the medical literature confirmed my hypothesis that the loss of bone density is also sometimes associated with Turner syndrome. While it is not possible to know the extent to which my bone structure and healing is compromised in comparison to women without Turner syndrome, I believe my dental history strongly emphasizes the importance of bone and dental care for women with TS.

I trained myself to ask questions of my doctors and nurses and always keep my appointments. It is extremely important to be your own best advocate, and be sure your questions are answered completely. We need to take extra special care of our hearts, with echocardiograms, regular monitoring of blood pressure and cholesterol, as well as bones, teeth and the thyroid gland.

I have learned to pick and choose when and how to tell friends and family about my diagnosis and talk about my treatment. No matter how you say it, it always seems somewhat random and out-of-the- blue to tell someone. But there are special people in my life who deserve that personal information, as well as special moments for me when it has helped immensely to open up to friends and family.

Turner syndrome inspires my curiosity, love of biology, and trust in God. The deletion of a piece of my second X chromosome in some of my cells is a foundation for biological curiosity and faith. Our Church calls each of us to "embrace the story you were meant to live". This directive truly expresses the scientific inspiration and faith that my diagnosis aroused in me, which continue to grow with my scientific research. Girls and women with Turner syndrome are beautiful and strong as well as unique. Everyone should embrace her own genetic make-up, and the talent that special combination creates in heart, body and mind.

Peter Pan and Me

Miriam Beit-Aharon

As a child, I believed that adulthood was only a matter of greater experience, plus a mysterious interest in the opposite sex. Despite our early ideas about adulthood, everyone must grow up. But I had a choice.

I was born a miniscule baby. A few weeks after my birth, you could see all my bones through the skin. My parents worried constantly, but somehow, month after month, I survived. Though I was far below the growth charts in height and weight and had a susceptibility to painful ear infections, I was otherwise healthy, proportional, but tiny. I grew very tired of hearing strangers call me "such a peanut" and other patronizing diminutives, but that was the way life was. I was aware that something was different about me, but then, no one is really normal. True, I was slow to learn certain things, such as the days of the week and how to read. But despite my difficulties, I was nonetheless happy and energetic. My life was good. I had a wonderful family and friends. I loved singing, photography, and gymnastics, where small size is an advantage. The dishes and silverware were kept in low drawers and cabinets where I could reach them. I never grew out of my favorite clothes, and I was comfortable with myself and with my capabilities, because I changed so very gradually from year to year. For my parents, however, this was cause for concern: what sort of kid doesn't grow out of her clothes?

When I was nine and a half, and only three-foot-eight, an endocrinologist finally ordered the karyotype that led to my diagnosis. Having been spared the obvious symptoms besides my size, I had been difficult to diagnose without a karyotype. There can be so many causes for failure to grow. It turned out that I, like ninety-eight percent of TS girls, was also going to need pills of the adult hormones my body failed to produce. The verdict: I would have to have growth hormone shots and then take estrogen pills when it was time to induce puberty.

Knowledge of my condition changed my perspective and I started to think of myself as a kind of mutant and began identifying with the X-Men. I now had to come to terms with getting a daily hormone injection and a blood test twice a year to monitor my dosage. The first time my mother came into the room with the syringe, I screamed bloody murder. I could just barely see the tiny needle at the end, but it was enough to help me work myself into hysteria. As soon as the needle had pierced my skin, however, I discovered that it really didn't hurt much.

My two older brothers played a major role in adjusting to this new reality. I will always remember when the younger of the two, who was thirteen, said: "It will be like this: you brush your teeth, you get your shot, you go to bed." He was trying to convince me that I was going to get used to the shots. It was very wise, as routine is exactly what the shot became. It was not painful, and I was never going to be upset about showing my friends who I was as a TS girl. Shots were something that I had to do every day, and my friends knew about it. My mother gave a very good shot, and altogether it was not the disaster that I thought it would be.

Still, I was not sure I wanted to grow. I was not used to it. The first time I grew out of a favorite dress was shocking, and I had very low spirits for the rest of the day. I was reconciled somewhat when people stopped making the aggravating remarks about my size, but soon height wasn't enough: my peers began to grow in a different way. One by one, my friends started talking to me about getting their periods, assuming I was getting mine too. They often wondered which boys I was interested in. I'm sure they must have felt that something was amiss. They would tell jokes whose meanings were beyond me, and as I did not share their new interests, we found that we had nothing to talk about. I tried finding new friends in children considerably younger than myself, but I could not truly relate to them either: they were too immature and too inexperienced. It was not their fault, but I could not blame myself either. My old friends, I felt, were the real traitors.

Despite the incentive to join my friends, when it came time for the pills, it was a struggle. Though almost fifteen, I had the mind of a child; I felt that *ick* factor so commonly expressed by children on seeing people kiss. I felt I had a choice of whether to grow up or not, and I was thinking: "no". A friend of the family described me as an old child, and I proudly compared myself to Peter Pan. "All kids grow up" I told my mother, "except for me." Though I understood how unnatural (in a natural sort of way) and strange that was, I could not believe that growing up was a better idea. It would mean destroying the body in which I had grown so comfortable. I could not comprehend being a teenager. I was convinced that I would unwillingly become a stereotype: jerky to my family, vapid with my friends, a boy-crazy airhead. I might not object to womanhood, but I could not imagine myself as a woman. Being a kid defined who I was. It took much persuasion for me to resign myself to the necessity of puberty. For the first few months I became physically ill at the thought of what I was doing, shuddering as the awful, sweet pill went down my throat. My brothers were very instrumental here too, convincing me that I didn't have an option, and couldn't hold my mouth shut to keep the pills out forever. They tried to reassure me that it would not be as bad as I feared.

Even after all the cajoling, and after beginning to feel the emotional benefits of growing up, I still could not look at myself without feeling slightly queasy. My induced puberty was smoother than its normal counterpart, and I was never taken by surprise, yet I pursed my lips in dismay every time I noticed a new change. Only while I was a junior in high school did I begin to accept that my body was simply doing what all girls' bodies do. My mind, too, developed in ways that I would not have thought to be hormone related. I grasped language in a new way—abstractions were suddenly comprehensible. Relationships had more depth than I had noticed before, and I could value people's virtues far beyond their ability to share in games. In a strange way, my memories changed too. I am now able to remember more, and remember better than I used to. All of the memories of times before I started taking the pills are like black and white photographs.

    I reconnected with my old friends, and they were happy to welcome me back. Despite this, high school was not a happy time for me socially. Having my old friends back saved my sanity on many occasions, but I had very few friends in school. However, one of my moments of triumph happened on the first day of my freshman year. I was walking down a back hall and saw a group of boys. As I passed them—the only person walking through—I heard them laugh. I was in a first-day-of-school mode and was not going to be pushed around already, so I turned around and said, "Are you wondering what an eleven year old is doing in high school?" I had estimated how old I looked, by my height and young appearance. In reply to my question, they asked, "because you're smarter than us?" in a really disheartened and pitiable way. Since they looked so sad, I had mercy and responded, "Because I am fourteen." They turned out to be very sweet, saying, "Sorry, we hope we didn't offend you!" I mumbled something about forgiving them and people sometimes not being what they appear to be. I could then walk on. Just because you are having a slightly low point in your life, it does not mean it isn't worth experiencing. It's all in the way you look at things, and situations usually get better from there.

    How lucky I am to have the brothers I have! They have helped me throughout my life, dealing with TS and in general. Noah always had the right thing to say. He made a joke when I had to relax a bit, encouraged me when I was nervous, and had good advice for me on what my outlook should be in order to maintain a good opinion of myself and accept myself as I am. Nathan is extremely important to me too, but in other ways. He is almost eight years older than I am yet has always shown me that he takes me, as well as my opinions, very seriously.

It is hard to accept that you are infertile, and I think that learning this at the same time as my parents did, and before the estrogen treatments, was important. Parents should not hide this information. A girl should know as soon as possible. I don't know anyone with Turner who disagrees with me on this point. I might not have understood it back then as I do now and may have even forgotten from time to time, but having it in the back of one's mind from childhood makes a big difference. It also brings about a greater confidence in and closeness to one's parents. I knew that I could always count on my parents to tell me what I needed to know.

Since I am only twenty years old, I still view getting pregnant negatively. Naturally, I had absolutely no envy for my freshman college roommate when she got pregnant in our first semester. But it did make me think, "wow, that could never happen to me, and I might never have a child of my own even when I want to."

There is still adoption and egg donation, so the future is not as bleak as it seems. Many women without TS are infertile, and sometimes go through invasive, deeply personal medical tests and trials just to find out what we know from our TS diagnosis. My parents gave me examples of some people like this that they know when we discussed the topic. The way I deal with the idea of infertility at the moment, as I can't help thinking and reacting to it, is by channeling my thoughts of children into enjoying other people's kids and playing with them. Each of these issues make us stronger and deeper, no matter how unnecessary and unpleasant each might seem to be at first. I find that I just have to recognize an adventure when I'm in one, and get through it with a smile and a lesson. I have had as easy a life as is probably possible in this world, and every day that I maintain this view is brighter than the last.

Who Am I?

Caroline Skene

How do we work through the process of finding out who we are and when does it start? What event kickstarts the questions and attempts to find out the answers? For me it began when I was diagnosed with Turner syndrome. I was just about to turn fifteen and had already made certain clear observations about *me*. I knew there was some difference between my peers and myself but frustratingly could not say what that was. Yes, I was shorter. Yes, perhaps I was sensitive and sometimes took things to heart too readily. I had insecurities like any other adolescent. But wasn't the process of passing through this stage and learning to understand your own internal voice what growing up was all about?

Many of the women I know with TS have exceptional talents. Some are artistic or musical. But it can require the devil-may-care attitude of maturity before the confidence to fully explore those talents truly surfaces.

Sitting in a consultant's room in a hospital in Edinburgh, Scotland, in the space of half an hour, so much of my future was placed in front of me. I now had an explanation for my past. The frequent ear, nose and throat problems I experienced as a child and I had almost forgotten now made sense. I now understood why, at fifteen, I still had the appearance of a girl four or five years younger. My relief at the time was palpable. Uncertainty was transformed into a medical condition with a name and a treatment plan. Ahead of me lay puberty under the control of synthetic hormones.

This was the beginning of a series of important physical and emotional changes. I learned that I would never have children who were genetically my own. However I was only fifteen, and the question of having babies was definitely not at the forefront of my mind. At least I could perhaps catch up with those around me at school. Therefore, on diagnosis, so much of the unease of the previous few years disappeared. I would take whatever pills the doctors gave me and everything would be right with the world.

The belief that the HRT tablets would be the panacea was of course naïve and a bit misguided. Nevertheless, the treatments increased my height by a crucial four inches. The knowledge that there was a reason for my short stature and delay in entering puberty in some way lessened their impact and, perversely, increased my confidence. I still remember the kindly consultant speaking to me as an intelligent patient and not as a child. He even ensured that he chatted to me separately, away from my parents, which I know was not the experience of the majority of women my age at the time.

I went through the remainder of my school years and then university with very little thought of Turner, beyond the consideration of "who do I tell?" I told a couple of close friends I trusted, but no one else. It was not something I felt necessary to broadcast. If I had a hunch it would influence a friendship or relationship, I would deal with the issue of disclosure at that point. I have always used gut instinct for making the decision whether to tell someone, and my judgment on that basis has never felt wrong. I cannot control the other person's response or who they tell, so I just take that into account before mentioning anything about Turner. Many people are satisfied knowing that you have a medical condition and no further questions are asked. Others may be more probing.

It was not until my mother found a letter relating to Turner in the problem page of a women's magazine that I had to ask myself how I felt about meeting others with TS. I was twenty-five years old by that time and living and working in Edinburgh, Scotland. The problem page led to contact details for a growth charity, which led to communication with the mother of a young girl with TS who ran a support group in Glasgow. I sat in the back of my parents' car full of trepidation as we drove along the M8 towards the hospital where the meetings were held. Questions raced through my mind. Do I really want to meet other women with Turner? What would they look like? Would I like them? Would I feel comfortable with them? I hoped so. The most prominent question was: "Will everyone look like me?" An illogical and ridiculous question, but the lack of clear information beyond a list of symptoms and characteristics in the medical texts allowed my imagination to fill in the gaps.

Thankfully, I found the mother to be warm and welcoming, and her daughter to be a boisterous, energetic, and endearing girl. I also met a woman who was the same age as I was, and while we had some obvious similarities and a lot in common, I was reassured by the fact that we had extremely different personalities. I had not lost my individuality. We became good friends and travelled to my first national conference together a few months later.

I have attended a large number of conferences over the years and have been lucky enough to attend international conferences in Toronto and Sydney as well as at Coventry in the UK. I won't deny there have been occasions when I returned from these events with mixed feelings. I did not want to associate my life with the difficult issues that had been discussed; yet I understood their deep relevance to me. I would see parents fearful for their daughter's future, their anxiety apparent in the tense conversations. I would want to tell them their worries were unsubstantiated, but in reality no one has a crystal ball to see what lies ahead. The most common recurring thought I had at the end of these conferences was a repeated affirmation that there was such a wide range in how this condition affects women. The needs of some were met with a light touch and maybe hormone treatment alone while others required more in depth intervention from healthcare professionals.

About twelve years after my original diagnosis I received a letter from a research team who wanted to reassess my karyotype for a study they were carrying out. The tests indicated I had a genetic profile that increases the risk of the streak ovaries developing a malignancy. Therefore to remove the risk, it was best to remove the tissue. I immediately accepted this as the correct medical decision, but I was not prepared for the emotional consequences of the surgery. Despite knowing and accepting that my ovaries were not functioning, it surprisingly came as a shock to realize that whatever had been there was now gone completely and could never be restored or revived. I imagined the surgeon attempting to locate the ovarian tissue within my body and cutting it away with a sharp scalpel. My psychological response was completely unexpected, and it took several weeks for the sense of loss to dissipate.

I am grateful to know I have a healthy heart, which is one of the more important medical concerns in Turner. I do not have a thyroid condition or diabetes. I do take calcium and bisphosphonates to improve my bone density. My HRT treatment has, with occasional minor exceptions, created no issues. Where Turner has had the most impact on my life is less tangible than echocardiograms and Dexa scans. It affects how I view relationships and how I view myself.

Like everybody else, I have strengths and weaknesses, and I respond emotionally to what is happening around me. I was overjoyed and excited when my married sister told me she was pregnant. I love spending time with my nephew. There are no pangs of jealousy when I see a woman with a baby. Conversely though, I do react to situations where I feel my worth is being devalued as I am a woman without children. I don't become angry or enraged, but shut down and become uncommunicative. I recognize this but have not yet been daring enough to confront those I feel are judging me on this basis. I do not know if I would have married and started a family had I been able to have children. My personal choice was not to go through the tough process of IVF or adoption.

The fact that I could not bear children has definitely influenced my perspective on relationships over the years. I was in a long-term relationship through my twenties and he knew about my condition from quite early on. We split sadly but amicably shortly after we both hit thirty with a significant factor being that we could not have our own children together. While painful at the time, it was a brave decision. Despite how much we cared for each other there were compromises that were impossible for him to make. The experience taught me the immense importance of strong communication, and as a result I have felt far more in control of subsequent relationships. It has made me less afraid to take risks and also appreciate a genuine connection with someone when I encounter it.

Not having children has also made me feel considerable pressure to achieve in my career. Rightly or wrongly, there is an assumption in my mind that my energies can solely be focused on my work, since I have not been juggling a full time job with the demands of bringing up children. While there may be practical reasons for enabling a more sustained career, it does not necessarily follow that you will be the big cheese in the boardroom as a result. Even in my forties I struggle to get the balance right in this area. It is important to be reasonably happy in the work you do. Not everyone can have their dream job but that is not an excuse to avoid aiming for something you enjoy doing, and having goals.

Where am I now, almost thirty years on from my diagnosis? Have I been successful in identifying who I am? I believe the journey of self-discovery and building self-confidence is a continual process. Being self-aware is a significant part of what is called emotional intelligence, an important life skill. I don't always get it right but I can use those not-so-great experiences to cope better next time.

I am currently at a stage in my life where some of the health issues, which luckily I have been able to dismiss so far, cannot be ignored. My hearing is deteriorating in one ear and I may have to submit to wearing a hearing aid in the not too distant future. For the last ten years or so I have enjoyed running to keep fit and maintain a healthy weight. Recently I have become noticeably slower and my left knee can give me pain from time to time. These may be minor problems and I certainly appreciate that other women have far more serious health issues to cope with (and I admire those who do cope) however these specific concerns are individual to me and I know my own body's capabilities and limitations.

    I will also have to make the decision on when to stop my HRT in the next few years. I feel the only way to prevent these hurdles from developing all out of proportion, and increasing in height so much as to be too high to jump over, is attitude combined with acceptance. If a hearing aid will allow me to pick up on and contribute to conversations that I may otherwise be tempted to withdraw from, then I will avoid the sense of exclusion that would arise if I don't have one. If I change my gym routine to a lower impact activity that is easier on the joints I will still be able to maintain an exercise regime. If I accept that aging is a natural process, in the same way as I accepted the Turner syndrome diagnosis all those years ago in that consultant's room, then I will continue to feel comfortable with myself.

Mountain Ways

Nadine Chaluck

Life can be difficult, especially when you feel different. I was diagnosed with Turner syndrome at age four. I've learned more about TS and what it means for me one step at a time. At first it meant growth hormone injections from age eight until age fifteen, when my growth plates fused. As I got older, it meant taking estrogen and progesterone pills. My final height is four feet seven inches.

My story about dealing with Turner syndrome is one filled with highlights and lowlights, black holes and gorgeous stars that twinkle bright. One star in my life is my mother. Of course, she knows I have Turner syndrome, yet she does not treat me differently from my siblings, and she is not shy about helping me with my problems. My dad, siblings, and most other people treat me as I want to be treated: as if there's nothing unusual about me. I wouldn't call it denial, it is just that I do not want to be constantly reminded that I have Turner syndrome.

I live in the small town of Golden, British Columbia. What is life like for me in my town? Golden is a small ski town in the Canadian Rockies, with about four thousand permanent residents. Mountains surround us. The ski hill, Kicking Horse Mountain resort and the Dawn Mountain Nordic trails are a gorgeous winter wonderland. I love cross-country skiing and I ski every weekend I can. Golden is quiet, and we locals are never in a rush. Everyone knows one another. Downtown Golden is calm, and full of a combination of old and older buildings that have been where they are ever since I can remember, many of them are unchanged.

Dealing with TS in a small town can be hard because there are no Turner chapters and my specialist doctors are three hours away in Calgary. It can also be easy, because everyone has known you since kindergarten and respects you. The few rude ones got all their teasing out of their systems in elementary school. For the most part, I have nice classmates and enjoy being with them.

I do talk about it with my close friends. I also usually explain Turner syndrome to the people who bug me about my short stature—and guess what? They usually never tease me about it again. I don't really enjoy explaining TS, though, because it's too complicated for some of my classmates. I did a lot of research, so I am capable of talking about it in detail, but some of my peers have a hard time understanding, which can be frustrating. Patience always helps. Through rumors and my short stature, most people guessed I have something. My method of deciding whom to tell to seems to be working out well for me. I only care if my close friends know details. I don't need all of Golden knowing my private business.

I do not remember my first injection, but I know it was my sister who held a paper towel on my leg and put pressure on it so that the syringe didn't bruise me for most of my first injections. I remember getting my first needle with the EasyPod in the Alberta Children's Hospital, as the nurse was teaching my Mom and me how to do it. I remember being nervous because a computer was about to give me an injection at the push of a button. The nurse told me I had to push that button. I was scared, so it took her a little while to talk me into it. It was a relief that it hardly hurt at all. After some time I got used to the idea and started giving myself my shots when I was twelve or thirteen.

If you can outweigh negatives with positives you have a good foundation for being happy and confident. The benefits are the Turner syndrome conferences. I have met so many really nice ladies that have Turner syndrome, and they are all an E-mail, letter, or phone call away whenever I need a chat. In so many ways for so many reasons, they let you know you don't have to go through anything, even Turner syndrome on your own. I contact my Turner syndrome friends all the time, so I never feel alone. The conferences themselves are so much fun, and they always include a cool outing. One year I went to one in Vancouver and we went to the Vancouver Science Centre. Another year the conference was at the Fantasy Land Hotel, West Edmonton Mall.

In recent years I have gone to Camp Carmen, held at Zajac Ranch camp for children. Zajac Ranch is summer camp on Stave Lake in Mission, British Colombia. Every week of the summer they have a different medical needs group, and Turner syndrome is one of them. Camp Carmen and other camps like it, in addition to the Turner conferences, are great for meeting and making new friends. Everyone is going through what you are going through and understand you completely. It is wonderful hanging around other ladies with Turner syndrome for a whole week.

Camp Carmen continues to be one of the highlights of my year. There are awesome counselors, activities, and games. It is so much fun singing, laughing, and just having a good time with amazing people you only get to see a few times a year. Zajac Ranch is where I first kayaked, where I first did high roping, where I rock climbed for only the second time, and where I first rode a horse. I did not know I was brave enough to climb to the top the high rope, but I learned there that I was!

Going through elementary school I did get teased a bit for being small. What I have found helpful in dealing with teasing are, a sense of humour and acceptance. When someone gave me a hard time about being short, I tended to come back with things like, "tell me something I don't know" or "so what". I dislike it when people ask me how old I am, especially adults, but if you accept who you are then everyone else will too. Self-confidence is so important. The bullying stopped because I did not let it get to me. I laughed and joked with the teaser and remembered that I have friends and family that love me simply for who I am. Me.

Golden Secondary (Grades 8-12), is a regional school with four hundred twenty students. I sometimes felt intimidated walking down the hallways of my school and standing in the concession line, because I am usually so much shorter than even the eighth graders. I cannot imagine being in an even larger school. My school life has been pretty successful, I think.

High school is going well, and I have made it through grade eleven. Many little adaptations have been made for me to make life easier. I have rungs on my chairs so that my feet don't have to dangle all day. Systems have been put in some of the classrooms for me and other kids who have hearing problems. There is no shortage of help available for people who need it—you just need to know how to ask. I study hard in school and that has paid off, because I am on the honour roll.

I have two siblings, a little sister and a little brother and have always had a good relationship with both. My sister, Dacey, is fourteen years old. We have some similar interests like movies and books that we love talking about. My brother Matthew is nine, and we're as close as a teenage girl can be with a fourth grade boy. I never fail to have fun watching movies and playing board games with my sister and brother. My Dad works far away from home, but I love hanging out with him when he's around. He is a structures foreman on a bridge project in Fort McMurray. It's hard having a Dad who "works away" as we say in Golden, but my family life is great. I think that having an awesome supportive family has helped me develop in so many ways; dealing with problems and creating and building self-confidence.

I have only recently come to complete acceptance about who I am and what I have to deal with. It was hard for me to develop the self-esteem I have now because I felt, and still frequently feel, that my parents and siblings cannot possibly grasp what I'm really going through. I have a close relationship with my Mom and tell her about a lot of problems. Having a good relationship with her has always helped me cope. She can comfort me and I love her even though I sometimes feel like my problems and thoughts are something that she wouldn't understand completely. How could she? She doesn't have Turner syndrome. While I do not think my mom understands everything I am going through and feeling, I do think that she has come as close as possible, which is pretty close. I have read a lot about Turner syndrome in order to get some different insights on the things I think are important to consider, and to get my questions answered. I read *Turner Syndrome Across the Life Span*, published by the Turner syndrome Society of Canada, which is an excellent book. It has everything you need to know about Turner syndrome in a format that is easy to understand. That book answered most of my medical questions. I also read *Growing up Small* by Kate Phifer. Kate is a funny and entertaining author. The book is older and a bit out of date, but the principles and advice are still valid.

Don't let bullying or stigmas stop you from doing anything you want to do. There is no shortage of help and good advice out there in the big world. Those are the life lessons I've learned so far. Now I would like to conclude with a favourite quote by Anais Nin:

"The personal life deeply lived always expands into truths beyond itself."

Rejecting Denial and Finding Myself

Jennifer Liu-Mormile

I am convinced I am here on earth to be an advocate for women with Turner syndrome. Currently, I speak at high schools to help educate students and biology teachers. TS is not the end of the world. It is your choice to be positive. I am Jennifer. I have TS, but TS doesn't define who I am.

Growing up as second of four children, I felt the world was my oyster. My childhood was happy, and I embraced life. I didn't go through any traumatic events as a young child and had good self-esteem. When I looked in the mirror at twelve, I saw a twelve year old, though my appearance was much younger. I thought, "So what if the other kids my age are taller, it's not that much." My sister is only thirteen months older, yet by second grade she was a head taller.

When I was a child, nobody treated me as if I had a condition, because I wasn't diagnosed until my late teens. At thirteen, I thought I was like any other teenager waiting for my period. I read *Are You There God, It's Me Margaret*, by Judy Blume, a popular book from the 1970s about a teenager changing into a woman with the coming of her period. I thought maybe I was a late bloomer, but my period never came.

I remember stepping on bamboo sticks on the arches of my feet to stimulate growth. My parents told me to try this. It must be a Chinese folk treatment. I also tried to drink a lot of milk. I never liked drinking milk. In fact, I hate milk, but I was willing to try anything. Because I am Chinese, and our characteristics include short stature, young appearance, and small bones (prone to Osteoporosis), these issues can be magnified by TS. Finally my sister or my aunt said something to my parents: "You have to find out what is wrong!" At fifteen and sixteen I looked the age, and was the height of my youngest brother, who was seven at the time.

My parents finally took me to Cornell Medical Center in New York City when I was sixteen in 1979 or 1980. Tests, including a karyotype, confirmed Turner syndrome. The next year was a whirlwind of activity that went by in a blur. I was in too much shock to understand what was happening, though the shock probably helped me get through it. Overwhelmed, we let the doctors do whatever they wanted without question. I even had to pose for a picture in just my underpants, with my arms out. I recall asking to have my chest taped. Thinking back, I was so naked and exposed. They use these pictures in textbooks. Attractive, isn't it? At the time I was so stunned that posing for the picture wasn't even a trauma.

I was treated with HRT and androgen. I just took that pill without asking questions or really knowing what it would do. My final height is four feet nine and a half inches, not much shorter than my mother or sister. Before treatment, the height difference was substantial.

My parents grew up during the Depression in Taiwan, which was then a Third World country. When my grandparents raised my parents they were interested in where the next meal was coming from, not their children's emotional state. Maybe that is why they never spoke to me about TS or how I felt. It seemed like a big secret, so I perceived Chinese culture as cold. Unfortunately, this is the only way my parents knew: the way they were raised themselves.

When I was six, my father had an opportunity to come to the United States for the Taiwanese government. What would my mother have done if she had known about TS when she was pregnant? I dread to think that she might have had an abortion. I have been lucky not to have any major complications. If I had, would my parents have been forced to take a harder look? They will have to give up denial when they are ready.

Starting college was a turning point. Looking more my age, I did not want to be different anymore. But I needed to come back to New York every three months to monitor the

HRT. I decided to go to a local gynecologist instead. This felt better. Every woman goes to a gyn. I had not had boyfriends yet, probably because until college I looked and felt like I was ten. My girlfriends were all very focused on boys, but I struggled with dating.

After graduation I met this wonderful man named Roy at a friend's party. He was of Irish, German, and Italian descent. My parents didn't mind. My sister married an Italian, and one brother married an Irish woman. Only my youngest brother married someone Chinese.

We were together for six years before getting married in 1993. Roy did not know I had TS. My parents never discussed how to tell anyone (especially the opposite sex), so I just ignored it, deciding it was OK to keep this secret. I told him I could not have any children but did not say why. But I wanted to get married, have the white picket fence, and have a family.

My mom took Roy aside and asked when we were going to have children, because we weren't getting any younger. When he reported this I thought, "She was sitting right next to me when the doctors said I couldn't have any!" I cannot be sure, but I do not think it was a language barrier and that she didn't understand.

When I was in my early thirties my friends and siblings were having children, while I was living in a world of magical thinking, imagining a stork would come one day and hand me a baby. Though genuinely happy for my siblings and friends, I was hurting inside. Someone tried to make me feel better by saying I wasn't missing anything by not having children. They meant well, but how could anyone say that? There will always be emptiness in my heart for the children I could never have. The die was cast when I was conceived, leaving me feeling robbed.

As long as Roy's youngest brother had no children, I coped. The turning point came when his wife finally had a son when he was forty, and I was the same age. We were asked to be godparents, but the baptism was at a Catholic church, and I was not Catholic. They wanted me to sign papers. Not feeling comfortable signing something that wasn't true, I asked my sister-in-law what to do. She said: "You shouldn't be worried about that, you should be more concerned that the Godparent has to be at least sixteen. You look like your are fourteen!"

I finally took a hard look at why I didn't have children. I am grateful that my brother in law had a child. What would have happened to me if I were still in denial, neglecting my health?

I decided to look on the Internet at work and learn more about TS. I was anxiety ridden and kept thinking: "Oh my gosh! Look at all these medical conditions!" I was a basket case, but needed to talk to someone, so I called my brother. He listened in stunned silence. This was the first real discussion I had ever had about TS,—at age forty, with my brother! The secret had been more important than possible medical complications. The next thing I did was find a cardiologist.

How had I lived with Roy all these years without feeling guilty for not telling him? I worried he would leave me. What would happen if the test results showed I need heart surgery? The hospital wanted an MRI of my heart. I was doing this all by myself. Racked with anxiety, I could not handle an MRI, so I had a CT scan instead. All the feelings about the doctors' visits at diagnosis were coming back. Waiting for the CT, I was terrified.

Panicked, I argued with myself constantly. "Some day I'll focus on TS, but for now I must be silent!" Followed by, "Keeping this huge secret from Roy is wrong ..." I could not take it anymore. Unplanned, I woke up one Saturday morning at eight thirty and just got it off my chest. He only said: "Is that it?" He had noticed that I hadn't been myself for a while. Later, he said that from what he read online, TS seemed pretty bad. He never got upset that I had kept it from him. He knew I didn't do it to spite him, and I thanked him for not seeing or treating me differently. I am grateful I finally told him.

The next year, at forty-one, I met another woman with TS. It was a profound, emotional moment that I unknowingly had been waiting for all my life. For the first time I didn't feel alone. Two years later, in 2006 I was ready for a national conference. It was daunting and overwhelming, with feelings of relief, joy, surprise and terror. Learning all the medical information was hard, but expected. I met wonderful people who will be my friends for the rest of my life. So far, there have not been other Chinese women at the conferences. I wonder where they are.

Before the national conference I told my parents where I was going. My mother said: "I don't want you to tell anyone in the company OK? I don't want you to lose your job". Keeping secrets again... I just said, "No mom, I'm not." I dropped the conversation before we got into an argument. My father has always been silent and distant, taking a back seat and not offering any opinions on any of us. Now I talk about TS openly, and my family is adjusting somewhat. You used to see in their faces how uncomfortable they were.

Meanwhile, Roy was struggling with bipolar disorder. Our situations were similar in some ways, with our parents in denial and not offering support. There was shame associated with discussing either condition. Never in my wildest dreams did I think I would marry a man with a mental illness, or imagine its role in my life. If you have Turner syndrome and neglect your medical care, (heart, etc) it could be fatal. With bi-polar, you can become suicidal.

Since "coming out" I still don't feel I have resolved all of the emotions that come with the implications of a TS diagnosis. I think that accepting a serious medical condition can be mourned just as when a loved one passes. But I was too busy coping with the much more serious matter of my husband's illness.

Roy committed suicide in July of 2008 He couldn't accept the seriousness of his condition, didn't take his medications, and was ashamed. The struggles and tragedy of the past few years put Turner syndrome into perspective, and made issues of infertility, short stature, and not being taken seriously because I look young, seem like nothing. What are they compared to coming to terms with my husband's tragic death?

My wonderful friends, mainly from high school and college, were there through thick and thin, helping me through this terrible period. I decided I owed it to them to tell them about TS. To my surprise and delight, they took it really well.

The following national Turner syndrome conference was only a week after Roy passed. I decided to go. I cried and I laughed. The loss was so great but the distraction so comforting I needed lots of people around me and kept thinking: "please let this conference last a year." It helped so much. The next conference fell on the first anniversary of his death, and has played a vital role in my healing.

At the conference they discussed TS as having a disability. I never saw myself as having a disability. When I was in school, there were no special tests to determine if you had a nonverbal learning disorder or any other special learning needs because of Turner syndrome. In fact, I had never heard of nonverbal learning disorder. I had no difficulties in school. It is good to have explanations for why we may have trouble in certain areas, but it could become an excuse not to try. For example, I had difficulty driving. It took until the third attempt to pass the road test. Had I known there was an association between driving difficulties and TS, I might have given up—but I didn't know, so I kept at it until I passed.

My college major was accounting! It is said that there is a connection between TS and difficulty learning math. This was a shock, because working with numbers is my profession. I may not have chosen that kind of work had I known. This is why I question emphasizing difficulties. Are we setting limits that do not have to be?

I recently had minor surgery. The gynecologist and cardiologist were fascinated by TS and completely forgot to be polite: "You take estrogen to make you look womanly. I just had a TS patient recently. She had breasts. Oh yeah, so do you." It is daunting to go to medical professionals who are so insensitive.

In 2009 the company where I worked as an accountant for over fifteen years moved the jobs from the New York office back to the UK, and let us all go. I felt I had lost everything. But I now have a much better job, and my life is improving. The moral of the story is I am still standing. Since the day I was conceived I have been a survivor.

In addition to advocating for women with Turner syndrome, I also advocate on behalf of people with Mental Illness. I have come a long way!

Damaged Goods or Specialty Item?

Jessica Esau

One day I was walking along the beach when I stumbled across an oyster shell. Most people would probably walk past and take no notice of it; after all, oysters are such plain-looking shells. But if someone takes the time to open it up and look inside, they may discover a pearl, a hidden treasure. Some people believe pearls to be the tears of the gods, but in fact, pearls are formed when an irritant enters the shell and becomes refined over time. My life is like that pearl.

I was born on August twenty fourth, 1992, at quarter after eight in the morning. By the standards of some, I was an ugly baby. Weighing in at a little over four pounds, I was a runt. My ears were misshapen and stuck out, extra folds of skin hung loosely from my neck, and one of my eyes was red and swollen shut. My appearance was affected because I have Turner syndrome. There are people would have considered me damaged goods that should have been discarded.

From a young age my parents taught me that God doesn't make mistakes and that he loves me and accepts me exactly as I am. In the church community I have found a place to belong and have discovered true friends. I've also had the opportunity to connect with other women and girls with TS through the local Turner syndrome chapter. This has allowed me to develop meaningful friendships with people who face many of the same obstacles that I do.

Sometimes I do wonder if my life would be so much better without Turner syndrome. It would certainly be nice to be able to reach things out of the cupboard without having to get a stool. I would love it if my feet could always touch the floor when I sit in a chair instead of dangling awkwardly in the air. It would also be a lot easier to learn to drive if I didn't have to choose between reaching the pedals and seeing out of the windshield. I would gladly be rid of these daily frustrations, and yet I realize that the challenges that have come from living with Turner syndrome have shaped who I am as a person.

My parents received the diagnosis when I was still in the womb. As they had never heard of TS before, they tried to gather as much information about it as they could. What they learned was devastating for them. In one of the first magazine articles they read, a mother who had aborted a baby with Turner syndrome wrote: "The tragedy started when our baby was conceived with a defect. A happy ending was never a possibility." I feel sadness when I think about that mother who aborted her child with TS. She missed out on the opportunity to grow as a person when she did not give the pearl a chance to form.

On that warm August day when my parents held me in their arms for the very first time, they saw a beautiful baby girl, a specialty item of infinite value.

My early years were much like those of any other young girl. I enjoyed looking at books, doing arts and crafts, playing with Barbies, singing along with Barney, and dressing up. Except for my regular visits to Children's Hospital, I was just like everyone else. Every now and then a kid on the playground would ask, "What's wrong with her eye?" My mom would give a brief explanation, and we'd continue playing.

For the most part, young children are very accepting of differences, and when they're curious about something, they are not afraid to ask. Adults, on the other hand, not wanting to be rude, don't ask questions and often make wrong assumptions. For example, one day while waiting outside the kindergarten room, my mom overheard some other parents discussing why I was so short. One mother said that it was because I had fetal alcohol syndrome! Many times through the years, adults have also sometimes mistaken me for someone with Down syndrome.

Because I'm only four feet seven inches tall, people often think I'm younger than I am. It's very frustrating to be spoken to in that condescending tone reserved for ten year olds. It's hard to be taken seriously for a job interview when you can barely see over the counter. However, despite all the negatives, there are some advantages. Getting charged the child rate at the movie theatre is always a sweet deal.

Although I was well accepted in primary school, the rules changed when I entered the tween years. In order to really fit in, one has to be pretty, slender and model the latest fashions and trendiest brand names. Most people would agree that adolescence is a trying time, but when you have a physical disability, it becomes that much harder. I shed many tears over being ignored on the playground and excluded from social get-togethers. Eating lunch alone and being the last one picked for group projects was also a bitter pill to swallow.

Rejection hurts, but it has made me stronger. Being on the outside, I have been able to see the world differently. I have developed empathy, insight, and compassion for the marginalized. Peer pressure doesn't have a grip on me, because I've learned to be confident in who I am. Many people view those with disabilities as an ugly- looking shell, not worthy of attention. As a result, they miss out on the beauty and joy inside that so many people bring to the world.

A Song In My Heart

Patricia Anne Selway

My name is Pat, I am fifty-three years old, and I am a woman who was born with Turner syndrome. I am five feet tall, was infertile during the childbearing years, and have an underactive thyroid, moderate to significant hearing loss, wide elbows, puffy feet, lots of moles, and early osteoporosis in my spine. Yet I consider myself lucky.

I have seven very lovely stepchildren: Paul, Luke, Joel, Alethea, Levi, Celestia, and Mia. At the time of my marriage in 1994, they ranged in age from six to twenty-one, with the oldest two already married and living on their own. I now have sixteen beautiful step grandchildren!

Born an unusually small baby, I played and grew up a very little girl. Being short posed no problems for me. I have a clear memory of a teacher at school, when I was perhaps six years old, asking us as a class what we wished to be when we grew up. Some boys answered that they wished to be firemen or doctors; some girls answered that they would like to be teachers or nurses, and I clearly answered that I wanted to be a mummy. I will add here that I was blessed with an extraordinary mother, and I could want nothing better than to follow in her footsteps. She has, in a humble way, defined womanhood for me, and has been a rock in my life.

The first real concerns with my health arose at fourteen. I had always had many friends, but I am also of normal intelligence, and it was not rocket science to guess that something was happening to my friends, and that I was somehow physically being left behind.

If I felt, at that time, that I had no one to confide in about my suspicions, the primary reason would be shame. I had a feeling that something was not right with my body, but in my case, that particular something somehow attacked the core of my feminine identity. I didn't understand what it was, but I clearly remember speaking to no one—not my parents, teachers or even my closest friends—because I was so ashamed of not being complete, or fully there.

When I was fifteen, a female nurse visited our school to conduct a check on our development, and I seized the chance to confide in her. She confirmed that I was a little bit behind, but she didn't see anything to worry her, and I will never forget the way she closed our conversation summing me up as a normal healthy girl. Case closed!

My astute mother tackled the subject with me when I was sixteen, taking me to the doctor. The correct diagnosis was due at least partially to a fortunate accident. The doctor had a relative with Turner syndrome, so he recognised my condition straight away.

When the consultant booked me in for the tests that would confirm my condition, he closed with the words "I'll tell you now, we think we know what it is; and there is only a fifty-fifty chance of you ever having children." (Not true, of course.)

Fertility and sexuality go together for most women, and for someone who has no such physical problems, I think that it is not possible to comprehend the emotional impact that infertility has upon a woman; it can bring deep shame, insecurity, fear, and grief.

My parents divorced when I was nine years old, and I had no contact with my biological father after the age of fourteen, which was before my diagnosis. My stepdad worked to help mum understand that she wasn't to blame, and tried to help me to be realistic about what the physical ramifications would be in my life, which I appreciated.

Throughout my life I wanted to be a mummy so much! Giving birth to children was at the centre of the role of a woman to my understanding, and it had been confirmed that I would not be able to do so. Not being able to fulfil what is by many thought to be a woman's most sacred function was sometimes very hard to bear. I can have so much love and sorrow in my heart at the same time. I still grieve for the children I have not borne.

My condition called into question my very identity. I was told that XY = boy, XX = girl, and I was XO.[4]

The diagnosis was a tremendous blow to my ego. I felt ashamed. Incomplete. I felt like a monster, an object of ridicule, and definitely of no worth as a woman. If I was not a woman, (and I knew for sure that I was not a man!) then what was I? An *it??* I felt so unlovable.

I was frightened about the future. What would happen to me? Would I gradually turn into a man? To some people such anxieties may seem ridiculous, but for me they were very real at the time. Would I die young? The consultant had told my mother that I would not hit forty. The fears themselves added to my shame, but to whom could I confide them?

In April of 1993, when I was thirty-two, my girlfriend from church called me and excitedly told me that she had found the ideal man for me. He was widowed, active in the church, and the father of seven. I very politely told her what I thought of her scheming, and nothing more was said.

This friend was responsible for organising a weekend convention of single adults in our area in August, and knowing my love for music, she pleaded with me to show up and organise a choir for her. I very reluctantly agreed though I didn't really want to go. I was happy to relax in my little flat of a weekend, but decided to attend after all. At the opening event I was kind of struck by a particularly handsome young man, carrying himself tall and straight. He stood out for his courtesy and gentleness. He was forty-seven, but he looked at least ten years younger. (He still does!)

We danced one dance that Friday night, and I was happy, but that was it. On the Saturday morning, when I was conducting my choir, the handsome young man walked in half

---

[4] Classic Turner syndrome was known as 45, XO. It has been reclassified as 45, X to end confusion over the use of O. See the glossary for further explanation.

an hour late, sat down, and started singing. I was stuck! Never in my life have I stared at a man the way I stared at him, I just knew I was looking into the face of the man who was going to be my husband. My friend was in the choir. She later admitted that she was dying of laughter because she knew whom I was staring at! We danced and got on well at the Saturday night dinner dance. At the end of the evening when I said goodnight to my girlfriend, she asked me if I knew who he was. I didn't! She clued me in: "Widowed, father, active in the church... Think about it Pat! He had no idea who you were, and you had no idea who he was, but you met and clicked anyway! YES!!" She laughed. I wore her wedding dress seven months later.

I finally decided to track down my biological father in 1997. We now have a fabulous relationship. When we first met up, he and my stepmother came to visit us, and the morning before he left, I told them about having Turner syndrome. He was silent, but my stepmother was wonderfully sympathetic and loving. A few days later I got a letter from my dad, who obviously took time to think about it seriously before commenting. He expressed his sorrow for me, and encouraged me very much in my relationship with my stepchildren. My stepmother has two children from her first marriage, so he has had the experience of stepparent-hood!

Oestrogen replacement, puberty, and menopause are very practical and physical issues that I am pleased to say my body had dealt with very successfully. I began oestrogen therapy at sixteen and so started artificial periods, and my body responded well to this treatment. I believe that I have enjoyed a measure of health that many women may not have, and consider myself fortunate. I was placed on hormone replacement therapy when I was thirty-seven and stayed on that for five years.

At fifty-two, I have two small fibroids, and a trace of osteoporosis in my spine, but no other problems that have developed with age. I have very sympathetic doctors who have given me assurances that I will age normally. This has been confirmed by my research on the Internet. Provided I take care of myself, strive to keep my weight and cholesterol down, and have my blood pressure and thyroid function monitored regularly, I should be able to look forward to a basically normal life span.

I have also made a personal commitment to lowering my meat intake, and refraining from smoking and drinking alcohol. Other steps include walking, yoga, and t'ai chi. I understand that I have an above-average risk of diabetes, stroke, and heart disease, and these are the main things I try to guard against. Confident about the future as I age, I am grateful for the core of girlfriends who surrounded me at the time of my diagnosis, encouraging and comforting me. They have been rocks in my life ever since. I am grateful for my Christian faith, a wonderful mother and a dear husband—all of these have provided the support and grounding that has been indispensable to me.

The main effect of living with Turner syndrome has been emotional. There is much that I do not take for granted particularly the level of health and the life that I have. I believe that I am more loving and compassionate than I would have been otherwise. TS has made me more aware of the emotional suffering that many people go through privately. I always aspire to be kinder, gentler, and more enthusiastic about life.

Clinicians that deal with the physical treatment and management of Turner syndrome should not overlook the emotional needs of young women who may be newly diagnosed, or ignore the unseen emotional impact that TS has upon a woman. Counseling can be beneficial, so please offer it. I know that at fifteen, having someone wise and professional to speak with on an intimate and emotional level, and without fear of judgement, would have made adjusting very much easier.

To young women I want to say, face the future without fear. Take care of your health, surround yourself with love and support, and educate yourself so that your individual questions and concerns are answered. Make a deliberate decision to live a full life, one that is satisfying and complete, leaving no space empty. I want young women with Turner syndrome to know that they can have a wonderful life. I am living proof.

Growing Up I Never Felt Different

Stefania Dimaio

It is exhilarating for me to write about my experiences as a young woman with Turner syndrome. I want to be a part of reaching out to others and sharing aspects of us that many people may not understand.

I was diagnosed at four, but I wasn't told until I was eleven. My younger sister and I were always treated equally. It's funny how as little kids I was bigger, but now she stands five inches taller. She has always respected me as her big sister and never gave me a hard time about size. At one point I longed to be taller, but there are more important things to focus on. Of course, in our family I never felt particularly short, because we are all tiny people anyway. I fit right in and always have.

TS is not something that makes me disabled; it's a part of me that makes me unique but never stops me. I am a mosaic, which in my case means that the karyotype showed that fifty percent of my cells were affected.

When the TS subject came up at one of my regular checkups with our pediatrician, Dr. Stockman, it was a reality check for my mom. She realized it was time to tell me and bring me to a pediatric endocrinologist. She felt I would never forgive her if she just let me go through my life untreated.

Learning of my diagnosis did change me, but the treatment I received from my doctors was all that I could ever ask for. They were always caring and loving. I owe a lot to Dr. Misra, and moving to adult care isn't the same.

At eleven, of course I was petrified of needles, but I learned to snap out of that fear soon after starting daily growth hormone shots! There were many doctors' visits to do blood work, but it really did become routine.

Though my parents saw me as perfect and whole, as I got older it became difficult for me to relate to the other girls my age. Middle school was brutal, as it is for many people. Of course I felt that my parents didn't understand how living with a genetic disorder affects me and that they were in denial. There were times when I had body image issues and low self-esteem. This troubled me, but like most young teens I kept that to myself. When these feelings became overwhelming, I tried to stay focused on the positive. Every now and then I felt like I was in a little kid's body but with the mindset of someone very mature.

Sometimes it surprises me that I am well adjusted, but because of the kind of person I am, and was raised to be, I was able to stay on a good path. While those around me followed trends and fads, or experimented, I always chose not to. At the time, I understood that these choices would not make me popular, but that was OK with me. Popularity is not always a positive sign.

Being part of such a loving supportive family always helped. They were the only ones who saw the side of me that was hurt by comments made at school, and they were there to talk me through the issues and reassure me. I am a strong person today because of these close relationships.

My mother and father are both from Sicily. Growing up, we had opportunities to enjoy the luxury of visiting our relatives overseas regularly. I loved these trips, which included plane rides and getting to know the people and places that are important to our parents. Our home was full of our strong cultural identity, with lots of expressions of love and caring.

Until age three I only spoke Sicilian Italian, which we still speak at home. When I started daycare, I learned English quickly. I am glad to have grown up bilingual and have a natural love and knack for learning languages. Using this skill whenever possible, I catch on quickly. Chinese and Arabic are fascinations of mine. It's a shame that budget cuts have limited the exposure of American students to foreign languages, because I feel it is important to know more than one. At Gloucester High School I took honors Italian all four years. It was my only honors class, and I felt very appreciated by the teacher, Signore Basile.

In high school I blended in much more. My height was no longer of interest, and my peers started to see my personality more than my height. I also became obsessed with heels and was never caught dead without them. My school social life was relatively normal. I was a good girl, not one to start or cause trouble, and well liked by a lot of peers and teachers, though I wouldn't have called myself incredibly popular either. Mathematics presented a huge challenge for me, but in the end I finished high school along with every one of my classmates. I know my parents were proud.

After graduation I did not yet know what I wanted to do. My boyfriend was a year behind me in school. We had been friends for a long time before we started dating, and our first date was his senior prom. Being young, of course I thought we would be together forever. I enrolled at Norwich University in Vermont, where my boyfriend wanted to go. Norwich is a private military college, but I was a civilian student. There were many positive things about the school and being in Vermont, but it was not the right place for me. I attended for a couple of years, but after we broke up I left. I realized that I wanted a trade and moved back to Gloucester. My parents were happy about my decision, and love that I am now close to home. I have grown and learned, and at twenty-three, I make the most of just being me.

Changes

Jordan Brilhante

When I was seven years old, I was diagnosed with Turner syndrome. At that time the only person I ever felt I could talk to about it was my grandpa. He was, without a doubt, the greatest man I will ever know. He was kind, sweet, gentle, and loving. However, above all, he was so understanding. He just had this wonderful way of listening to me and seeing my point of view through any situation. He was my go-to person always, whether or not it had to do with TS. I was fourteen years old when I lost him to congestive heart failure. After he died, I felt unsure of who I was, losing myself through the pain of his death. Things were just not the same without him.

As a result, I started high school feeling even more outside myself and distant from others. It was as though no one could really hear me or was listening to me when I spoke. That summer, my mother signed me up for a Turner syndrome camp at Pepperdine University. It was the greatest experience of my life up till that point. Suddenly, there wasn't just one person to go to, or relate to, but fifty! It was a feeling of joy that will never be forgotten or lost. These girls have truly changed me for the better. I am not trapped inside myself anymore. Before, life was very dull and not very exciting, filled with average people, everyday issues, and typical high school melodrama. Nowadays, it's not like that anymore, and not just because I am now in college. I can call on these girls whenever I need or want support, regardless of the circumstance. I am there for them, and they are there for me in ways no one else could ever imagine, though we do not live physically close to each other.

Looking back and remembering the loneliness, and thinking about all the trouble I got myself into for no good reason, it seems ages and ages ago. At times, it even feels like it was happening to a completely different person. Relationships were difficult to maintain before I met my Turner syndrome chicks (as I like to call them). Making friends used to be so tough to do. Sometimes it seemed it wasn't worth all the trouble. Before the experience of TS camp, I think many people found it hard to be around me. I had a reputation for being controlling or manipulative. I never really believed that I was, or at least I never was that way on purpose. I must just have expressed my personality in a way that made people feel that I wanted to intimidate or manipulate them. Now, all thanks to my TS friends, I don't feel I express that rather demanding quality as much as I used to do. Also, I don't worry about being alone anymore.

At times it seemed even my family did not understand who I really was or the motivations behind my personality and behavior. I did not truly appreciate their situation or issues either. Turner syndrome had caused my family a lot of problems, as well as being a source of heartache. It was a financial drain for my mom and dad and a struggle for the other children in the family. It was hard for them to put up with all the extra attention I received. They might not want to admit it, but it is true. I think this is why my older brothers and sister sometimes teased me a lot and even gave me the very harsh nickname of "Chromo." But after returning home from camp, things were different, and our relationship improved quite a bit. I could embrace Turner syndrome for what it is and how it is a part of me, and who I am. No one can understand me the way my girls do, and they have helped me feel more secure, and made me less needy at home.

Before meeting them at camp, I used to ask God why? "Why was I chosen to be the one in two thousand or so girls affected with Turner syndrome? Of the approximately sixty to seventy five thousand women and girls with TS in the USA, God, why did I have to be one of them?" Previously, I liked to think he left this question unanswered. But then at camp, I felt there was a response. Turner syndrome was a gift, that provided the opportunity to meet these most amazing people. I used to see TS as a curse from all sorts of different angles, but now I realize that it also has advantages.

    Also, Turner syndrome is no longer an excuse for my behavior. This genetic mutation or however anyone wants to call it, is a part of me and always will be. I was born with it and I will die with it. There can be no simpler explanation than that. My religious faith has made me stronger through it all and wiser. I now really enjoy life for what it is, and for all the good even Turner syndrome has brought me.

The Circle of My Life

Susan Lazar

"All my life's a circle; Sunrise and sundown; The moon rolls thru the nighttime; Till the daybreak comes around."
– Harry Chapin, *Circle*

    I am fifty years old. My journey with Turner has certainly not been a straight line ; more like a circle. The most important discovery on this journey? My life has been richer by living it as a woman with Turner. I am stronger, unique, and much more empathic. We all have different life experiences and characteristics that make us who we are. Being born with Turner syndrome has certainly shaped who I am, not just physically but also as a person.

    I was first diagnosed with TS at age twelve. My parents found a wonderful endocrinologist who was both sympathetic and matter-of-fact about my condition. At the time, I processed the news to mean that I finally had an explanation for why I was shorter than all of my friends. Also, I knew why I had no breast development and had not yet gotten my period. I was very happy when estrogen took care of puberty. In other words, Turner didn't mean all that much to me. Yet.

    As I got older and went away to college, I was not unhappy that I didn't have to worry about birth control. But always lurking in the back of my mind was the understanding that I would be unable to become pregnant when I wanted to, when I was ready to build my family. In my late teens I also began to get a fuller picture of Turner and how it might affect my cognitive functioning, as well as contribute to some obsessive-compulsive thinking and social anxiety.

    In my mid-twenties I met the man who was to become my husband. I shared with him that I had Turner syndrome after watching the movie *Raising Arizona*, about a couple who are unable to have children. As our relationship became more serious, we talked about my infertility and what it would mean for our future. My husband turned to his parents for advice; thankfully, they told him to follow his heart. They assured him that if we loved each other, we would find a way to build a family. I will always be grateful for this. So with the support of both of our families, we married, and soon thereafter we adopted first a beautiful baby girl, and a year and a half later a wonderful baby boy.

There remained other issues. As fortunate as my husband and I have been, and as happily as I was raising my family, I struggled with jealousy as my three sisters and many friends became pregnant and increased their families. Of course, my sisters understood and were supportive. I also found it helped to explain TS to my friends, and they too were very understanding. I simply wished, and probably still wish, that I could experience the miracle of pregnancy. I also wished that my husband and I could have decided whether we wanted more children without worrying about the financial burden and emotional turmoil of the adoption process.

In the end however, I feel immensely blessed. I am grateful for my parents and three sisters who never saw me nor treated me as different from themselves. I am thankful that my parents made sure I received the best possible medical care, and for the terrific endocrinologist who took care of me. I am grateful for my husband and my in-laws who always accepted me exactly as I am.

Today I have an internist who takes care of the physical difficulties associated with Turner (osteoporosis, high blood pressure), and I am lucky to have her. I also have been seeing a psychotherapist for many years who has been instrumental in guiding me on my journey towards self-acceptance. It is important for everyone to take care of his or her physical and mental health. How much more so for a woman presented with the challenges that Turner brings?

As a psychotherapist with Turner syndrome, I constantly wonder how many of my personal challenges are connected to Turner, and what difficulties I share with other Turner women. As I have aged, and especially after many years of my own personal analysis, I have realized that I have denied that Turner syndrome really had anything to do with *me*, or the core of my being. Of course Turner syndrome deeply affected my sense of myself as a complete woman and a competent, productive person. I wondered if these feelings were only my own, or if any were shared with my Turner sisters. Now that I have had some experience interacting with young girls and adolescents in a group setting, as well as much time talking to other women, I know that we share these feelings of otherness, of being different. Finally, hopefully, we arrive at a place where we can embrace who we are, Turner and all, knowing that we are worthy and lovable with our own strengths.

Then there is the whole sex thing. We may feel inhibited; we may feel not completely female (we had to take estrogen to go through puberty, after all) and this can affect our sex lives. Unless we talk about it and face the impact TS has on our sexual psyche, we will never fully enjoy the sex lives to which we are entitled. We should be free to enjoy our partners (gay or straight), orgasms, and everything in between. But for this to happen, we must embrace and love our very selves.

So, my journey with Turner has indeed been a circle. I started out as a young teenager who neither understood nor appreciate the full complexity of Turner. My life as a young person did not feel much impacted by this condition. As I got older, I began to face the physical impact as well as the emotional impact. Now, at fifty years old, Turner again doesn't feel like it has much of an affect on my life. That is the circle. This time, however, it is a feeling born of a fuller understanding and appreciation, and not just of Turner syndrome. It is an understanding of the challenges I have faced and of the people who have helped me to face those challenges. It is a deep, deep appreciation of the people who love me. And in the end, isn't that what really matters?

Appendices

Xo

Appendix 1

Discussion Topics

by Susan Lazar, Psychotherapist

Adult Women Discussion Group
Let's talk about Turner syndrome and our sexuality

Moderator:

Before separating participants into groups, have each person introduce herself and tell the group something she would like everyone else to know. They can then divide into pairs or small groups, depending on the number of participants.
Explain to the participants that everything that is said in the room must stay in the room. In order for everyone to grow and learn together, there must be respect and trust. Distribute the questions below to the participants to discuss in their respective groups or pairs.

Alternatively:
* Pick just one or two questions.
* Give different groups a different selection of questions.
* Put all the questions into a box or a hat[5] and have different groups draw a question at random.
* Have each group choose one person to report back to the assembly later, followed by an open discussion.

Discussion Group Questions:

• How do you feel your parents dealt with your diagnosis? Were they overprotective? Were they comfortable talking to you about body issues, including infertility?

• Did you get age-appropriate information growing up? If you were unable to or did not want to talk to your parents, was there someone else you could go to, such as a close friend, doctor, teacher, therapist, or religious counselor?

---

[5] Choosing a whimsical, creative hat may be a good ice breaker.

• The teenage years are a difficult period for many people. Do you think Turner syndrome made your experience different from that of other awkward teens, and how? Do you think you worried about your appearance more or less than your peers who did not have Turner syndrome?

• What coping strategies did you use?

• What do you like about your body? (your hair, eyes, breasts, legs)

• What parts of your personality do you particularly feel are your strengths?

• What about sexual development? How has Turner syndrome affected your relationships?

• Do you believe you have been able to enjoy a sex life to your own satisfaction?

• Do you think that having Turner syndrome in some ways inhibits you, and prevents you from having or enjoying sex?

• For women with Turner syndrome infertility can be a most painful emotional experience. Let's talk about what it has meant to you, and how it has affected your view of yourself as a sexual being.

• Let's talk about the full range of feelings we have had with regard to infertility and sexuality.

Teen Discussion Group
Let's talk about it.

Moderator:

Start by setting a tone of trust and confidentiality. The young women need to agree that these conversations will be strictly confidential. Begin by having each girl introduce herself; ask them to say something that they would like everyone else in the group to know about them. Depending on the number of girls, they can work in pairs or small groups. Give out the questions, and give the girls ten minutes to talk to one another. You can tell the participants that one person can give a report when you all come back together.

Alternatively:
* Pick just one or two questions
* Give different groups a different selection of questions.
* Put all the questions into a box or a hat[6], and have the different groups draw a question at random.

Teen Discussion Group Questions:

• When did you find out you have Turner syndrome (how long ago, and how old were you)?

• How were you told, and do you remember how you felt?

• How did those feelings change as you got older?

• Do you feel now that your body is different from other girls' bodies?

---

[6] Choosing a whimsical, creative hat may be a good ice breaker.

- Who can you talk to if you are worried about something regarding your body and sexuality?

- What do you like best about your body?

- What parts of your personality do you particularly feel are your strengths?

- How has your view of yourself, up to this point, been shaped by Turner syndrome?

- How do you feel about femininity?

- Has Turner syndrome affected your view of yourself as a girl, and as the woman you are becoming?

- What gives you confidence?

- Do you think that your peers are more comfortable talking about sex than you are?

- How do you feel about infertility at this point in your life?

- What questions do you have about Turner syndrome's effect on you that have never really been answered, particularly regarding sex and sexuality?

- Let's voice our hopes and fears about sex and relationships. No questions are out of bounds. Remember that everything we say stays in this room.

- What are your hopes, dreams, and fears for the future?

Parents Discussion Group
Parenting girls with Turner syndrome

Moderator:
Before separating participants into groups, have each person introduce himself or herself. Below are some suggested opening remarks:

• How old were you and how old was your daughter when she was diagnosed?
• How old was your daughter when she learned of her diagnosis and how long had you known?

Now the participants can divide into pairs or small groups, depending on the number involved. Parental couples should be in different groups.

Explain to the participants that everything that is said in the room must stay in the room. In order for everyone to grow and learn together, we have to trust one another. Distribute the questions to the participants and have them discuss various questions in their respective groups or pairs. Alternatively:

* Pick just one or two questions.
* Give different groups a different selection of questions.
* Put all the questions into a box or a hat[7] and have different groups draw from the box at random.

One person in each group can report back to everyone before having a general, open discussion.

---

[7] Choosing a whimsical, creative hat may be a good ice breaker.

Discussion Group Questions:

- Who told your daughter she has Turner syndrome, and how did you feel about her response?

- Had you ever heard of Turner syndrome before you found out that your daughter has it?

- Can you share your first thoughts at diagnosis?

- Were you previously suspicious that something was amiss?

- Did you get what you needed from medical professionals at diagnosis?

- Was your daughter treated with respect?

- How did you come to terms with your daughter's diagnosis so that you felt you were ready to help her?

- How do we encourage our daughters without making them feel that we are denying the serious implications of Turner syndrome?

- How do we protect our daughters' medical privacy without having them think that we are ashamed because they have Turner syndrome ?

- One of the hardest tasks for all parents is to let children make their own mistakes and accept their independence as they grow up. Can you share the strategies you have used to support your daughter's ambitions without taking away her autonomy?

- We each have a blind spot when it comes to our children and Turner syndrome. Can you identify an aspect of Turner syndrome that you shy away from?

Appendix 2

Karyoptyping in Plain English

Karyotyping in Plain English

After our daughter's diagnosis, we struggled with the meaning of her karyotype. Though extremely motivated to learn, we had no tools to interpret what we saw or to understand the jargon.

With the fear and frantic energy common in parents of a child diagnosed with a potentially serious condition, we were focused on fixing it. I wanted to repeat the test, hoping to get a better score. Maybe we could still get a grade of *normal*? We had always been suspicious that something was wrong. She had never grown as expected. But it was hard to accept these definitive results.

At first glance a karyotype looks strange. The matched pairs bend towards or away from each other at odd angles. Sometimes one of the pair is straight, the other crooked. This does not matter. It is the actual measurable length that matches each chromosome with its partner, and this is how they are identified. Their posture is no indication of their health. They look this way because the process of getting at the DNA in the harvested cells roughs them up a bit.

The karyotype is used for detecting a variety of genetic conditions. It does not show the intricacies of your genetic structure, it only allows you to eyeball the general size and shape of the chromosomes. What can be seen is whether a sizable chunk of a chromosome is missing,[8] repeating, or whether substantial pieces have switched places.

This test confirms Turner syndrome, but not which specific characteristics (phenotype) will be present. A karyotype with a lot of genetic material missing can lead to a relatively mild phenotype. One such example would be a woman who *only* has premature ovarian failure. Someone else's karyotype may show many normal cells and only a

---

[8] Missing? Where did this genetic material go? We do not know the answer.

small percentage of the cells with deletions, yet this individual might have more major physical issues.

How DNA variations in Turner syndrome lead to certain outcomes is still not well understood. Though more modern microscope technology is now used, the karyotype test was developed in the 1950s and it provides only raw information on the chromosome's structure. Nowadays we expect more detail, but as of this writing the karyotype is the tool that confirms absolutely the presence of Turner syndrome. The next step is a very thorough medical exam to see which symptoms are present.

When I first encountered my daughter's test results twelve years ago, I felt helpless. After devoting so much time to understanding it, how can I transmit this information to other people in the same situation? I have decided to go with a recipe format, because I love to cook, though this is *quite* a recipe. (Don't look for it anytime soon on the Food Network!) The people who do karyotypes are highly educated and trained, and the process takes about a week to complete.

Ingredients
• One vial of human blood from a single person or a small piece of skin or other tissue (if the subject is in utero, amniotic fluid or a bit of placenta).
• instant cell culture or special media to grow cells. (including hormones and nutrients.)
• separation solution
• instant "stop mitosis" chemical
• cell-bursting solution (to burst the cell wall without damaging the chromosomes)
• cell wash
• fluorescent dyes (to color the different parts of the chromosome - sometimes called G-banding)

Equipment
• light microscope
• microscope slides
• special digital camera
• specialized computer software

Method
    For chromosomes to be visible under the microscope, we need them in their most compact, natural form. This happens when they are actively dividing (mitosis), so we need to catch them in the act.
    First, separate the white blood cells using the separation solution. Discard the other blood contents. Only white cells are needed to observe the chromosomes. (Cells do find instant cell culture delicious, but it will take up to a week to grow the thirty to fifty cells needed).
    When the cells are just about ready to divide, the chromosomes are at their most compact, so add the "stop mitosis" chemical. In this recipe, timing is everything. The cell's processes are now frozen, with their information accessible.

Now, to see these compact chromosomes, get them out of the cells by putting them on a microscope slide and adding cell-bursting solution. After the cells burst, wash away the leftover debris. Voila! The chromosomes are now stuck (fixed) to the slide, and each cell's contents are in a small but unmatched bunch.

Since chromosomes have no color, put dye on the slide. Now the chromosomes look like short strings with light and dark bands, and we can tell them apart. The dark or light bands each include hundreds of different genes. It is time to take a digital photograph of our results and load that onto a computer.

Now arrange the dyed chromosomes attractively in pairs according to their sizes. Use the software carefully. This is labor intensive, because we have to manipulate the data even though the picture of the chromosomes is now on a computer. The pairs are called one to twenty-three, with pair number one being the longest. In each pair, one chromosome came from the mother and one from the father.

Carefully check for missing or additional material, or structural changes, like bits that could have switched places on the chromosome (translocations). Deletions, additions, mutations or translocations can occur on any of the chromosome pairs. Those changes on the sex chromosomes (pair 23), are the ones that identify variants of Turner syndrome, and that is what we are looking for.

Our karyotype test results now show the total number of chromosomes, and their structures can be seen. Now we can print a digital picture of the chromosomes, arranged by number, and we have a luscious copy of our results to serve. Bon appetit!

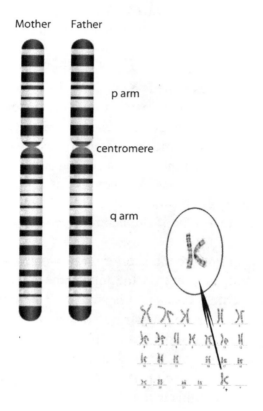

Figure 1: The drawing on the left shows a normal human X chromosome pair, one X inherited from the mother and one from the father. Pictured on the right are all 23 chromosome pairs of a normal human karyotype. Each band within the chromosome contains specific genetic information passed on from generation to generation. Each time a cell in our bodies divides, both maternal and paternal X chromosomes are first duplicated and then each of the new cells receives one maternal and one paternal copy. This is true of all 23 chromosome pairs.

Understanding the results

So we have our recipe complete, but now what? I took biology in high school but was not a science major in college. Please take heart and a deep breath as a math-phobic English major counts to forty-five (or forty-six), and describes the architecture of the twenty-three pairs of chromosomes in our DNA:

We focus on pair 23, the smallest pair, the sex chromosomes—nicknamed X and Y. Everyone needs one functioning X to survive. Two Xs indicate a woman, and XY indicates a man. One good X is enough for a woman too, but she does have Turner syndrome. Sometimes a woman will say she is XO[9] to indicate the missing second X chromosome. [10]

The narrow waist-like part, about a third of the way down on each chromosome, is called the centromere. The sections on either side of the centromere are known as the *p* and *q* arms.

The *p* arms are the shorter ones; when lined up for the digital portrait, they are on the top of the image. The *p* just stands for "petite," because it is smaller. The longer arms are called *q*, not because they resemble a *q* in any way but because *q* follows *p* in the Latin alphabet. At first blush, *p* and *q* might lead you to think that maybe some part of the chromosome is just facing the wrong way, which is what I thought at first, but this is not true.

Another possibility among many is that the second X chromosome is missing in only some of the sampled cells. This is called "mosaic." Perhaps one of the X chromosomes is minus either the *p* or the *q* section, or has two of one section or the other, instead of one of each.

---

[9] Classic Turner syndrome was known as 45, XO. It has now been reclassified as 45, X to end confusion over the use of O. See the glossary for further explanation.

[10] Missing? Where did this genetic material go? We do not know the answer.

Perhaps a small snippet of Y is attached to one of the Xs? In another case, a percentage of the sampled cells are missing an X, but even though the rest have XX, one of the members of these pairs has two *q* arms. All of these variants, and many more, signify Turner syndrome, but they do not tell us specific symptoms.

A cell has a "choice" between two versions of each gene. Each chromosome comes in a pair, so go ahead, cell, choose one! But how does this selection take place? We don't yet know for sure. The area on the edge of a cell might have a silencing effect on genes, with the center being more active, but this geography is not yet understood. The position of the chromosomes in living cells is not revealed by a karyotype. Some research has hinted at the possibility that the specific location of the chromosome inside a living cell might be responsible for which gene is expressed, or turned "on."[11] In TS there is often only one version of the X in many of the woman's cells, or one healthy one and one variant. If the cell uses the gene from an X where that gene is healthy, a particular feature or symptom of TS may not be present.

In other conditions an extra chromosome, or an extra part of a chromosome, is the main identifier. Many variations can occur on any of the different chromosome pairs, but not all of them are compatible with life.

Turner syndrome is seen on the karyotype, so it is genetic, but this does not mean it is inherited. Turner syndrome happens randomly across all ethnic and socio-economic groups all over the world at the same rate. Also, it does not occur more frequently in older parents.

---

[11] See Scientific American February 2011, p.71

Here are some common Turner syndrome karyotypes, and what the shorthand stands for:
- 45, XO or 45, X is common. The 'O' signifies where the second X chromosome should be. It does not represent the number zero. For a live birth to occur one fully functioning X chromosome is vital. The second X chromosome can be seen as a backup. However, some genes on the second X are active and are necessary for some female biological functions. It is not yet understood why genes are turned off or on in the second X. In recent years it has been concluded that no one alive is actually 100 percent 45, XO or 45,X. These women are now understood to be subtle mosaics.
- Another variant is called isochromosome Xq. An isochromosome is a chromosome that has lost one of its arms and replaced it with a copy of the arm it has. So a second X is there, but it has two repeating q segments rather than one p and one q. The p segment is gone completely, replaced by an extra q.
- Isochromosome Xp, with two p arms and no q, is also a possibility.
- Yet another variation is a "ring" chromosome, sometimes called a circular X. There, the ends of one of the X chromosomes in the cell fuse together to form a circle seen only at the time of karyotyping.

Which symptoms of Turner syndrome will be expressed? These cannot be predicted accurately by the karyotype, but only through physical examination of the person.

One symptom very common in Turner syndrome is short stature. This gene has been identified, and has been named "SHOX." which stands for "short stature homeobox" gene. Homeobox genes regulate development and are found on the sex chromosome (pair 23). Two copies of "SHOX" seem to be important for bone development and growth. If you have Turner syndrome and you are about the same height as other women in your family without the help of extra growth hormone, then your SHOX genes may both be present and functioning. The loss of one copy of this gene is considered likely to be the cause of the short stature so common in Turner syndrome. The gene responsible for premature ovarian failure or other specific symptoms of Turner syndrome have not been found as of this writing.

New research and discoveries on genes and how they are expressed are being published all the time. We should soon have a better understanding of their effect on women with Turner syndrome.

Most people haven't had the occasion to see their own karyotype, so they have never had the opportunity to think about their genes or chromosomes in this intimate way. That is changing. Having your genes sequenced, (and with more information than a karyotype), is becoming less expensive, and more and more people are having this sort of testing. Women with Turner syndrome will have the opportunity to watch the general population wrestle with many of the questions they had upon confronting their karyotype. In trying to make sense of this limited and confusing information, some people are tempted to identify their personal characteristics by the genetic profile in their karyotype. But this test cannot detect exactly how Turner syndrome will affect your health, or its impact on your mathematical ability, automotive prowess, career, or love life. Having the results of a karyotype does not make you any more a prisoner of your genes than anyone else, but for now you have extra information. All living creatures are a magical combination of genes hidden on those chromosomes, and the mystery continues.

Appendix 3

Hormones and Facts about Our Favorites

Hormones and Facts about Our Favorites

Here are a few of the roles that hormones play in the body:

* Regulating the metabolism
* Preparing the body to fight or flee
* Activating or inhibiting the immune system
* Stimulating puberty or menopause
* Generating hunger pangs
* Directing bones to grow to make children taller
* Motivating fat cells to get larger or smaller to adjust weight
* Controlling the calcium levels in the blood and bones

Hormones are chemicals that regulate many vital reactions, and they are not limited to growth, puberty, or mood swings! They don't perform actions themselves. They give a message to cells to tell *them* to get the job done. Hormones regulate many important functions in all living things larger than one cell—including plants. In people, they are produced in various glands, or sometimes even by individual cells.

Hormones use biochemistry to work their magic. There are tiny receptors for different hormones on or inside different sorts of cells. Each kind of hormone has targets on specific sets of cells. Growth hormone and insulin, for example, attach themselves to the cell surface. The hormone locates a port of just the right size and shape to deliver a chemical message to the cell. This contact generates a cascade of reactions, because the cells follow these instructions to the best of their ability. Some of our other popular hormones like estrogens and androgens (female and male hormones) enter the cell and then attach to specific chemicals inside, affecting the genetic machinery or DNA. If cells are given too many messages, the effect can be like screaming instead of asking nicely. Mistakes and over-reactions can ensue. For example, excess growth hormone can cause growth in the wrong proportions, extra large hands or feet, severe headaches, or even diabetes. The diabetes is temporary *if* the excessive growth hormone is reduced to the proper level. This is why growth hormone use is monitored so carefully. Great caution is required to deliver exactly the right amount so as not to produce negative side effects.

Insulin, produced in the pancreas, tells cells in the liver, muscle, and fat tissue to take up glucose from the blood. As people with diabetes can attest, the right amount of this hormone is extremely important to well-being.

The pituitary gland, which sits just under the brain, releases some of our high-profile hormones, governing growth and sexual development, and many others as well. It is sometimes called "the master gland" but it is itself controlled by a part of the brain called the hypothalamus. The pituitary is composed of two parts the anterior pituitary, which makes many important hormones, like growth hormone and those controlling puberty and the posterior pituitary, which acts as a storage site for hormones made in the hypothalamus. Together with the hypothalamus, the anterior pituitary orchestrates many of the hormone deployments in our bodies, even those that it does not produce itself. For example, the thyroid gland takes up iodine from our food and converts it into T3 and T4 to balance our metabolism, affecting every organ in the body. If the T3 or T4 levels in the blood are too low the pituitary and hypothalamus notice and the pituitary sends thyroid stimulating hormone (TSH) which instructs the thyroid to make more T3 and T4.

The pituitary is amazingly small, about the size of a shelled peanut. It weighs only about 0.035 ounces or half a gram. In people with Turner syndrome the pituitary is normal, excreting a normal amount of growth hormone. However, someone with Turner syndrome usually requires some extra growth hormone to boost reluctant cells to achieve. At the time of this writing, it is not clear exactly why the cells of a girl with Turner syndrome are more reluctant to listen to the *grow* message. The interaction of hormones with cells and with individual genetics is exquisitely complex.

Are the ovaries producing estrogen? The pituitary sends out follicle stimulating hormone (FSH) to direct the ovaries. This chemical prodding by the pituitary tells the ovaries to release estrogen (female hormone) into the bloodstream to generate signs of puberty and the other functions of estrogen. If the ovaries don't follow orders, the pituitary sends out even more FSH to encourage them. High levels of FSH can be measured fairly easily with a blood test to show if the ovaries are not working. Taking estrogen supplements causes the FSH levels to drop, because the pituitary stops nagging. All women going through menopause have high follicle stimulating hormone levels. Girls and women, whose ovaries are not functioning, including those with Turner syndrome if they are not taking estrogen, have high follicle stimulating hormone levels. There are girls with Turner syndrome who have normal ovarian function, at least for a while.

If an adolescent with Turner syndrome takes estrogen supplements, her uterus and breasts will grow and develop.

The birth control pill, consisting of artificial estrogens and progesterone has been in widespread use since the 1960s. Over that time it has been honed and calibrated to be safe and effective not only for controlling fertility, but also for treating a wide variety of conditions including endometriosis, polycystic ovarian syndrome (PCOS), and debilitating periods, all in women who don't have Turner syndrome. Since so many hundreds of millions of women have used the pill, the patch, or other estrogen and progesterone treatment options, women with Turner syndrome can safely reap the great health benefits they provide.

The Miracle of Growth Hormone (GH)

Hormones were first specifically identified and isolated (separated and distilled from the blood) beginning at the end of the nineteenth century. The pituitary had been recognized long before, but mostly due to tumors that were sometimes found on it.

It turns out that growth hormone is a very complicated chemical, even relative to many other hormones, and some of its functions are still a mystery. But from the perspective of the unique needs of a girl with Turner syndrome, growth hormone is most useful for two purposes:

* Stimulating the growth or development of every part of the body

* Helping retain calcium and strengthening bones (estrogens also contribute here)

Unlike estrogen, growth hormone was extremely difficult to make. The pill became available in the early 1960s, but it took another twenty years to produce growth hormone in the laboratory. Before the man-made variety became widely available in 1985, girls with Turner syndrome did not often receive growth hormone. Short stature as present in Turner syndrome was not considered a serious enough reason to experiment with this tricky hormone from natural sources. Why?

Growth hormone used to be obtained at autopsy from the pituitaries of people who had died, but serious unanticipated health risks ensued. In 1985 it was proved that preparations of growth hormone from human pituitaries could be contaminated with the proteins that cause Creutzfeldt-Jacob disease, a degenerative brain disease that has no cure, and is always fatal. Over one hundred people died from this worldwide.

Synthetic growth hormone is made in the laboratory by microorganisms that have been altered to contain the gene to make growth hormone. It has proved to be a wonderful aid for many children with growth problems, but it can be a dangerous drug in the wrong hands, so a tremendous amount of security surrounds its production and distribution. Professional athletes and wannabe athletes crave it in the belief that it will enhance their performance. Other people wrongly think that it is the secret to the fountain of youth, and that they will stay young and strong if they can fool their bodies with injections of growth hormone.

After almost thirty years, synthetic growth hormone still needs to be injected, while estrogens can be taken by mouth. Why? Estrogens and estrogen-like substances can be naturally absorbed in the stomach whereas growth hormone cannot. Unfortunately, digesting growth hormone does not work as a hormone delivery system. The good news is that when you have finished growing, the daily injections are finished too. In addition, the needle used to inject this tiny amount of growth hormone is very thin and fine, which minimizes the discomfort.

The more you learn about the many hormones your body uses all the time, the more amazed you will be by their almost magical properties.

Appendix 4:

Turner Syndrome
Frequently Asked Questions

Turner Syndrome
Frequently Asked Questions

What is a syndrome?
A syndrome is a group of symptoms that occur together, or a condition with a cluster of those symptoms. Some syndromes are genetic, but many are not, just as some genetic conditions are inherited, while many are not. What makes the condition a syndrome is the grouping of the features together in the same person, not the direct cause. Many syndromes were identified long ago, before the medical techniques to explain their source existed. The most recent example that comes to mind, totally unrelated to Turner syndrome, is Acquired Immune Deficiency Syndrome (AIDS), which was originally identified in the 1980s by its collection of symptoms. Later, it was understood to be the result of a virus.

Turner syndrome was first recognized in females by the grouping of shorter than-expected stature, delayed puberty, and infertility. In addition, it was often accompanied by additional characteristics such as a low hairline, wide carrying angle of the elbows, puffy hands and/or feet in infancy. Eventually, the "cause" of this collection of symptoms was discovered to be a random genetic alteration, though questions remain as to why this leads to many of the specific symptoms.

What is Turner syndrome?
Turner syndrome is a genetic condition in which one of the affected girl's two X chromosomes is either missing or deviates from the norm in at least one of many ways. The genetic material is lost when the deletion, which is random, occurs sometime after conception, during early cell division. Any couple can have a daughter with Turner syndrome.

What happens to the genetic material?
It is gone, reabsorbed by the body.

Doesn't *genetic* mean *inherited*? If Turner syndrome is genetic, doesn't that mean you get it from your parents?
Not all conditions that are genetic are inherited. Your genes can be affected by sources other than your parents' DNA. Turner syndrome is a random gene mutation that occurs at about the same rate in females from all ethnic, social, and economic backgrounds. The frequency is not affected by the age of the parents, as in some other chromosomal variations, such as Down syndrome, which happens more frequently with older mothers. Turner syndrome is caused by changes to the genes during early cell divisions of the embryo and is not inherited from anyone in your family. Yes, in extremely rare circumstances, a young woman who has Turner syndrome will go through puberty unassisted, and then ovulate and conceive naturally. There are two articles that discuss these very few documented cases of six women who gave birth to children with unremarkable genetic profiles, and five who gave birth to girls who also had Turner syndrome.[12] However, it is important to restate that if your mother did not have Turner syndrome, you did not inherit it from her. It just happened randomly.

Who was Turner?
Turner syndrome is named after an American endocrinologist from the University of Oklahoma. Dr. Henry Turner published an article in 1938 describing seven women patients with

---

[12] Muasher, S., Baramki, T. A. and Diggs, E. S. Obstet Gynecol. *Turner phenotype in mother and daughter.* Dec;56(6):752-6 (1980).

Portnoï, M. F., Chantot-Bastaraud, S., Christin-Maitre, S., Carbonne, B., Beaujard, M. P., Keren, B, Lévy, J., Dommergues, M., Cabrol, S., Hyon, C. and Siffroi J. P. Familial *Turner syndrome with an X;Y translocation mosaicism: implications for genetic counseling.* Eur J Med Genet. Nov;55 (11):635-40 (2012).

similar physical features. (including short stature, lack of post-pubertal sexual development, increased skin folds in the neck, and a wide carrying angle of the arms). In 1930 a German pediatrician, Otto Ullrich, described the same physical characteristics in female patients, so Turner syndrome is sometimes known as Ullrich-Turner syndrome. It is not clear why Turner's name is the one that stuck in North America. The actual chromosomal deficiency was not identified until 1959 when karyotyping was refined It was seen that one of the X chromosomes was missing or altered in women with the characteristics of Turner syndrome. Please see Appendix 2 on karyotyping for detailed information.

How many people have Turner syndrome?
The statistics vary somewhat, but Turner syndrome occurs internationally at a rate of from one out of 2,000 to one out of 3,000 live female births. Assuming a world population of seven billion, then there are probably anywhere from 2.3 million to 3.5 million women and girls living with Turner syndrome in the world today. Eight hundred new cases are diagnosed each year in the United States.

What are chromosomes?
Chromosomes are genetic material; each parent contributes twenty-three, making a total of forty-six chromosomes. One of the twenty-three pairs, the sex chromosomes, determine the sex of the baby, but have other genes on them as well. A male will usually have an X and a Y chromosome (46, XY) and a female two X chromosomes (46, XX). Girls and women with Turner syndrome often have only one X chromosome instead of the usual two in all or many of their cells (45, X) or (45, XO). The "O" represents the missing X chromosome. The loss of genetic material (X chromosome) occurs randomly during the cell divisions that follow conception. Represented as either 45, XO or 45, X, this is known as classic Turner syndrome. Sometimes the X is missing from only a portion of

the cells (46, XX/45, XO), commonly known as Turner mosaic. There are a number of other variations in the karyotyping for Turner syndrome, including ring chromosome. (see Appendix 2) Sometimes a small snippet of a Y chromosome may be present. This is known as mixed gonadal dysgenesis. A geneticist can explain your karyotype. Genetic counseling is often recommended for those diagnosed with Turner syndrome.

How is Turner syndrome diagnosed after infancy?
To be certain of the diagnosis, the doctor will draw some blood for a karyotype. Sometimes skin cells are used for the test instead. This test identifies the presence of a damaged or deleted X chromosome. See Appendix 2 on karyotyping.

What about prenatal Diagnosis?
Prenatal diagnosis is sometimes made using chorionic villus sampling, amniocentesis, or occasionally a sonogram. A sonogram, however, will detect only the more obvious signs indicating Turner syndrome, such as a bicuspid aortic valve.

When are girls diagnosed?
Most girls diagnosed in infancy have cardiac symptoms or swelling in the hands and feet, which should lead to immediate suspicions by medical personnel at the birth. If these better-known symptoms are not present, diagnosis might come later in childhood when growth almost stops, or in early adolescence when there is no growth spurt, and a lack of development of secondary sexual characteristics. Not infrequently these signs are missed, in spite of short stature, ear infections, etc, and young women can reach adulthood before diagnosis. All girls under the third percentile for height should be evaluated for Turner syndrome. The pediatrician, physician or nurse should also suspect Turner syndrome in any girl who is not showing signs of puberty at the appropriate age. A quick examination of the carrying angle of

her arms and/or of the arch of her palate can raise suspicions further, especially in conjunction with frequent ear infections, even in an otherwise healthy girl. Some women are never diagnosed.

Why use estrogen treatment?
If the ovaries of girls or women of menstruating age are not functioning for any reason including Turner syndrome, they need estrogen supplements for the sake of their general health. Estrogens have many other effects beyond maturing sexual characteristics, including bone strength and neurological and cognitive development. It is important for overall health to continue with estrogen supplements until the natural age of menopause.

Why use growth hormone treatment?
Growth hormone is effective in increasing by several inches the adult height that the average girl with Turner syndrome would otherwise reach. If a girl is growing on a curve that suggests that she will reach a height that is acceptable to her, she might not need this therapy, but most girls will need and want this help.
Girls with Turner syndrome have normal levels of growth hormone in their blood, and the reason for the short stature so often seen is not completely understood. A small boost in the form of daily growth hormone shots over a number of years is usually most effective. Careful monitoring of the dosages is very important to avoid unpleasant side effects. The shots can be administered till either the height goal is achieved, or x-rays show that the growth plates in the bones have fused and no more growth is possible.

What is hormone replacement therapy (HRT)?
HRT refers to the various estrogen and progesterone combinations given to women who cannot make enough of these hormones on their own. Most physicians try to imitate

the usual progression of puberty with low doses of estrogen either orally or in the form of a patch. This is important for girls with Turner syndrome because very low doses of estrogen work with growth hormone to increase height toward the end of the growth period. Doctors use various schedules to increase the estrogen gradually after that. When the young woman has enough estrogen in her body, progesterone is added for seven to fourteen days at the end of each cycle, usually monthly, in order to induce menstrual bleeding. Once her normal cycle is established, a young woman could stay on her HRT with a separate pill or patch for the estrogen, and a pill for the progesterone, but it is often easiest to switch to an oral contraceptive or birth control pill that contains both estrogen and progestin or to a patch that contains both. The pill or patch is available with estrogen and progesterone either alone or in combination. A doctor can help the individual decide which therapy is best for her. Of course it is important to remember that if you are sexually active, neither the pill, the patch, nor any other HRT protects against sexually transmitted diseases (STDs).

What exactly is the pill?
Birth control tablets contain either estrogen and progesterone or progesterone only. A girl with Turner syndrome should take a combination pill so that she gets estrogen to keep her bones healthy. One pill is taken every day, and it is best taken at the same time each day in order to become a part of the person's routine. For one week of the month the pills actually contain no hormones but just serve to make sure that the schedule is maintained. It is during that week of taking "empty" pills that menstruation occurs. For convenience some girls choose to take a pill preparation which induces a period every three months rather than every month. The pill, like all oral medicine or nutrients enter the bloodstream after first being processed by the liver. Your doctor will advise you if a hormone patch would be a better choice for you.

What is a hormone patch?
When given via a skin patch, estrogen and progesterone enter the bloodstream through the skin, without passing through the liver. A drawback is that direct sunlight or high heat affects the patch, and could cause the release of a big dose on a hot day, leaving less to release later. Those using the patch also need to stay away from electric blankets, hot tubs, or anything else that would heat it up. Tanning beds are bad for everyone, but even worse if you are using a patch.

Why have an internal exam (pelvic exam), and what does it entail?
If you are over nineteen, this can provide important reassurance and information about your female plumbing. Ask any questions about sex or sexuality. The information is held in strict confidence and the clinician has heard it all before! Explain if you have abnormal vaginal bleeding or secretions.

There are often pictures taped to the ceiling to help you relax because nobody adores this exam. Relax your mouth, jaw and stomach to relax the vaginal muscles. A lubricated speculum, which looks like a metal duck's bill, is opened in the vagina to see your cervix. This should not be painful. Vaginal walls stretch, but they need to be held open to take cell samples with a little brush or cotton swab and see what is going on. It only takes a moment. The clinician will also feel the size and shape of your uterus.

During a standard pelvic exam doctors do not check for sexually transmitted diseases (STDs) automatically. If you are sexually active you should be tested. Women with Turner syndrome, like all women, need to guard against STDs.

Can you really be 100 percent 45, XO (45, X)?
The current understanding is that women with Turner syndrome are all mosaic in varying degrees, and that one good X still needs a little help here and there for a baby to survive. Therefore no one is 100 percent 45, X (45, XO).

What can an individual who has Turner syndrome expect today?
With good medical care, an individual can look forward to a normal life expectancy and a good quality of life. She can maintain good health, since the serious medical conditions associated with Turner syndrome, if present, can be addressed.

Why are there so many treatments for issues in Turner syndrome?
Large numbers of people in the general population have conditions that occur in Turner syndrome. This means that many excellent and frequently used therapies are available. Some examples include: thyroid deficiency, estrogen deficiency, infertility, diabetes, ear infections, high blood pressure, coarctation of the aorta, and bicuspid valve disease. Good treatments exist, but it is up to individuals to make sure that they follow up and monitor their condition for good quality of life and in some cases, life itself.

What about ear and hearing issues in Turner syndrome?
Recurring middle ear infections (otis media) are one of the more common hallmarks of a childhood with Turner syndrome. Later, in adulthood, a very high percentage of women develop hearing problems. Interestingly, Dr. Henry Turner made no mention of hearing issues or ear infections in his description of the collection of symptoms commonly present in the syndrome. However, the vast majority of women with Turner syndrome will have issues with their ears. There are many hypotheses about why childhood ear infections are more frequent, severe, and longer lasting in girls with Turner syndrome than in other girls. These range from differences in the angle of the ear canal, low-set ears, the width of the passages inside the ear, or to lack of estrogens.

In addition to ear infections which can lead to hearing loss, there can also be cholesteatoma, a benign cyst behind the ear drum. It is very important for women with Turner syndrome to have their ears checked regularly to preserve hearing. Everyone should do their best to protect their ears from the insult of loud noise through headphones or ear buds. It is important to wear earplugs to reduce the noise level at concerts and in other loud environments. Hearing loss in Turner syndrome tends to strike the part of hearing in the middle range of sounds, making following conversations very tiring. Recent advances in hearing aid technology are impressive, and can help considerably.

What is coarctation of the aorta?
Coarctation means the aorta is narrow in a specific place on that major artery (the aorta), where it arcs downward. According to current medical opinion, this is because for some reason the blood supply to that specific area of the aorta was reduced during the development of the heart. A lack of adequate blood flow to this area can impede the development of the artery, and cause it to be too narrow in that spot. Even a very small disturbance of blood flow to the heart during development has major repercussions for that section of the aorta. Fortunately a surgery can repair this narrowing.

Coarctation of the aorta occurs in Turner syndrome with estimates varying between 5 to 20 percent. It also accounts for over five percent of congenital heart defects in the general population. Because of the relatively small number of women and girls with Turner syndrome and coarctation, perhaps your cardiologist has never had the opportunity to treat someone with Turner syndrome, though s/he may have seen many cases of coarctation. You may find yourself in the role of ambassador of the Turner syndrome community!

What is a bicuspid aortic valve?

The aortic valve is supposed to have three flaps (tricuspid.) If your aortic valve has only two flaps you have bicuspid aortic valve (BAV). If so, you need good follow-up and routine checkups with your cardiologist. Three flaps is a sturdier design, allowing for a tighter seal that prevents blood from flowing back in the wrong direction. Over time, a bicuspid aortic valve can also become stiff, and may not open as widely as it should. bicuspid aortic valves are also sometimes associated with structural abnormalities of the aortic wall itself, making the aortic wall weaker than it should be. This can make the aortic wall prone to a bulging (called an aneurysm) that can develop a tear in the lining of the aorta. This is called aortic dissection and is very dangerous. Some people with bicuspid aortic valve also have coarctation of the aorta, and there may be a connection between the two conditions. Not everyone with a bicuspid aortic valve needs a surgical repair immediately, because in spite of missing a flap, the valve can manage to work well for a long time. However, this condition should not be ignored, and must be monitored regularly. Bicuspid aortic valve occurs in up to two percent of the general population, so there is a lot of experience with bicuspid aortic valve in cardiology. There are good treatments available. For more information contact the Bicuspid Aortic Foundation.

Why is heart monitoring important for everyone with Turner syndrome?

Aortic dilatation and aortic dissection can occur, though very rarely, even in girls with Turner syndrome who do not have a cardiac problem. For this reason all girls and women with Turner syndrome should have cardiac monitoring at regular intervals.

What are horseshoe-shaped kidneys?
Horseshoe-shaped kidneys have this distinct appearance when the two kidneys fuse and grow together at the bottom during fetal development. They then form what looks like a single horseshoe shape. The ureters (tubes that carry urine from the kidney to the bladder) should be checked to see if they are creating back-flow into the kidneys because of their angle. Horseshoe-shaped kidneys also occur in one out of five hundred people in the general population. No treatment of the kidneys themselves is usually required, but this shape can increase the risk of high blood pressure as well as the severity of urinary tract infections, and should be monitored regularly by your doctor. Up to 15 percent of women with Turner syndrome can have this malformation of the kidneys.

What is hypothyroidism?
If your thyroid gland is under performing it will be seen on your regular blood test. The treatment is a pill of synthetic thyroid hormone taken daily on an empty stomach. This medication is not expensive and is easy to use. Hypothyroidism is more common in women than in men, and is extremely common in the general population.

What about motherhood?
Reproductive technologies are advancing quickly, and since female plumbing is intact in women with Turner syndrome, with the exception of the ovaries, pregnancy can be achieved with a donor egg and in vitro fertilization (IVF). However, there are significant risks which each woman with Turner syndrome must consider very carefully. At this writing there are several important considerations before a woman with Turner syndrome can consider a pregnancy:
1) She must have been taking replacement estrogens, probably from the earlier teenage years, (or have menstruated spontaneously), in order for her uterus to be the size needed to sustain a pregnancy to term.

2) It is vital to consult a cardiologist with experience in Turner syndrome, because pregnancy and labor can be extremely dangerous for any woman with Turner syndrome. It is even more dangerous if she has had coarctation of the aorta, and/or a bicuspid valve, even if they were repaired.

3) High blood pressure is a concern, as well as diabetes, so a thorough medical workup is extremely important.

What about sex and Turner syndrome?

Both data and anecdotal evidence suggest that many women with Turner syndrome find it challenging to explore sex and sexuality.

An American study reported that of eighty women with Turner syndrome, 55 percent had experienced sexual intercourse. The mean age at first intercourse was twenty-three.[13]

Another study published in France found 30 percent of the 566 six participants with Turner syndrome had had sexual intercourse by the age of twenty, versus 85 percent of women of the general population.[14]

There seems to be a consensus that women with Turner syndrome become sexually active later than the general population. Why? Attempts have been made to answer this question in terms of endocrinology, self-confidence, body image, education, infertility and gender identity. There are suggestions that the age when estrogen supplements begin might be a significant factor as well. There are discussion questions in Appendix 1 that explore this topic further. In addition, Diana Clifton's essay deals specifically with sex and sexuality.

---

[13] Pavlidis K., McCauley E., and Sybert V.P., *Psychosocial and sexual functioning in women with Turner syndrome.* Clinical Genetics(1995).

[14] Carel et al. *Self-esteem and Social adjustment in young women with Turner syndrome.* Journal of Endocrinology and Metabolism; 2006.

Is there a comprehensive list of all characteristics associated with Turner syndrome?

There are many lists that can be found both in medical texts and on the Internet, but no one individual has more than a handful (some more than others) of symptoms. No one reading those lists should worry that they can or will develop all or most of them. The most common signs of Turner syndrome are: extremely short stature compared to children the same age and other family members, absence of growth spurts at the expected times, plus the absence of most of the signs of puberty. Any girl who is below the third percentile in height should be checked for Turner syndrome so that any serious issues can be addressed in a timely manner.

Turner syndrome symptoms are often listed in a random way, mixing the serious with the cosmetic, a practice that can be confusing as well as terrifying. Please remember that you, or the woman or girl you love who has Turner syndrome does not have all these symptoms. Some are cosmetic, and some may not surface until middle age or beyond, but good medical care is essential in order to treat or monitor the more serious conditions, and minimize their impact.

Serious conditions that can occur with Turner syndrome and need to be monitored and addressed:
- Heart defects such as bicuspid valve or coarctation of the aorta, aortic dissection or dilatation
- Kidney/urinary tract problems
- High blood pressure (often related to the heart or kidney defects)
- Middle ear infections and related hearing problems
- Ovarian failure causing a lack of estrogen for pubertal development as well as infertility
- Osteoporosis, a risk for all women who have insufficient exposure to estrogen

- Autoimmune conditions such as type I diabetes, hypothyroidism (an under-active thyroid) or Hashimoto thyroiditis and arthritis

Are there cosmetic identifiers of Turner syndrome?
There are some cosmetic signs of Turner syndrome that are often overlooked by the general public, or even by a pediatrician or internist. However, the trained eye of an endocrinologist or another woman who has Turner syndrome might notice some of the following:
- Irregular rotation of wrist and/or elbow joints (cubitus valgus)
- Extra skin on the neck (webbed neck)
- Puffiness or swelling (lymphedema) of the hands and feet in early infancy
- Lower hairline at the back of the neck
- High arched palate (requires peering into the mouth to see)
- Small lower jaw
- Lower-set of the ears
- Relatively broad chest with wider spaced nipples
- Soft nails that turn upward a bit at the ends
- More moles than other members of the family
- Drooping eyelids
- Short fourth and/or fifth finger and toe on both hands and/or feet

Spatial awareness
Spatial awareness problems are often reported in Turner syndrome, and one of the diagnostic tests is to ask the girl to draw a hexagon. Inability to do this is considered a possible symptom of Turner syndrome. However, studies investigating hexagon-drawing abilities in the general population are not available for comparison.

What about Turner syndrome and math?
Connecting Turner syndrome directly with poor math skills is highly problematic, because many women have significant trouble with math. Some women who have Turner syndrome have excellent math skills, even as others struggle. For an interesting study on how a person's performance is affected by self- stereotyping, please read the following:
http://icos.groups.si.umich.edu/shihpaper.html.

Interestingly, this article shows how Asian-American women, when directed to think of their Asian heritage, do better at math problems than when they are directed to focus on being female! Another example explored in the study is the elderly. When they identify themselves principally as "old" they don't do as well on memory tasks than if another identifier is emphasized. This is not to say that individuals who have Turner syndrome do not struggle with spatial or math issues. However, awareness of the effect of expectations is important. Identity as a woman and a woman with Turner syndrome in particular, should not discourage anyone from trying her best. The results might be surprising.

Are there learning disabilities associated with Turner syndrome?
Learning disabilities that are associated with Turner syndrome are also widespread in the non-Turner syndrome population. Relatively common ones include non-verbal learning disorder, or difficulty with spatial concepts, or math, and/or a poor sense of direction. With professional help, as well as determination, such problems can be overcome or mitigated. A woman with Turner syndrome can live a confident life. Overcoming challenges is part of the human condition, and not unique to Turner syndrome.

Appendix 5

Recommended Reading

Recommended Reading

The Boston Women's Health Book Collective. *Our Bodies, Ourselves for the New Century: A Book by and for Wsomen*. Touchstone, New York, New York, USA, 1998.

This book started as a large pamphlet in 1971, and has gone through at least nine editions. It is the bible of women's health. This is an important book for every woman (fifteen and older) who does not yet have it on her shelf. It is a very comprehensive discussion of all women's health issues, as well as social issues like body image and emotional wellbeing.

The Boston Women's Health Book Collective. *The New Ourselves, Growing Older*. Touchstone, New York, New York, USA, 1998.

This book focuses on women's health and social issues at menopause and beyond.

Hamilton, Jill, and Irena Hozjan, Editors. *Turner Syndrome: Across the Lifespan*. Castlemore Graphics, Brampton, Ontario, Canada, 2008.

Published with the support of the Turner Syndrome Society of Canada, this is an excellent resource on medical issues in Turner syndrome. Thirty experts from different fields relevant to Turner syndrome contributed to this book.

Phifer, Kate Gilbert. *Growing Up Small: A Handbook for Short People*. Paul S. Eriksson, Middlebury, Vermont, USA, 1979.

Written by a woman who does not have Turner syndrome but is four feet nine inches tall, this book discusses the issues and ramifications of being short in the late 1970s, as well as the authors's own experiences growing up in the 1940s. It is somewhat dated, but that is also a point of interest, showing how far we have come in some respects. She has also authored a book on thoroughbred horse racing. I was disappointed that she did not include any insights from jockeys, who are known for being extremely short, though they are usually men.

Appendix 6

Recommeded Viewing

Recommended Viewing

If you would like to see some excellent online film explanations of karyotyping, cell division and DNA, here are a few. You might want to watch them more than once. They are wonderful tools for understanding these complicated processes.

Brightstorm, *DNA structure*
www.youtube.com/watch?v=tknYrU98rBk&feature=channel

Brightstorm, *Karyotye*
http://www.youtube.com/watch?v=q8errsrd4FE

Brighstorm, *Mitosis*
http://www.youtube.com/watch?v=zVDnrTTp1jw&NR=1

ppornelubio, *Mitosis*
http://www.youtube.com/watch?v=VlN7K1-9QB0&feature=related

Appendix 7

Support Organizations

Turner Syndrome Support Organizations

Turner Syndrome Foundation
www.turnersyndromefoundation.org
This organization is based in New Jersey, not far from New York City. They focus on increasing the understanding of Turner syndrome, and heightening the interest in scientific advancements. The foundation have social and educational events throughout the year.
Their seven short videos on YouTube called *Turner Syndrome Diaries,* are required viewing for anyone interested in Turner syndrome. They are beautifully composed and well edited professional interviews, with excellent production values.

Turner Syndrome Society of Canada
www.turnersyndrome.ca
This wonderful organization for Canadians interested in and/or affected by Turner syndrome, also welcomes visitors from south of the border!

Turner Syndrome Society of the United States
www.turnersyndrome.org
TSSUS is based in Texas, but has chapters in many parts of the United States. Go to the website to find a chapter near you! The website also features a list of more than twenty-five international Turner syndrome organizations and their contact information. The society's annual conference offers opportunities to learn about the latest medical therapies and discoveries, as well as social activities for girls and women with Turner syndrome of all ages and their families.

Turner Syndrome Support Society (UK)
www.tss.org.uk
This is the premier organization in the United Kingdom for practical information as well as contact and social opportunities in a warm, friendly, and supportive environment. The Society maintains contact with international groups and hosts and promotes gatherings for people affected by Turner syndrome in every region of the United Kingdom, as well as organizing an annual conference.

Other Support Organizations

There are many other advocacy and support groups who deal with almost any specific challenge you could be facing. Don't hesitate to reach out. You can meet wonderful people. Here are a few of these organizations that focus on issues that can also affect women with Turner syndrome:

The Arthritis Foundation
www.arthritis.org
The Foundation is one of many organizations, foundations and support groups because arthritis is such a common problem. They run a children's camp for those who have Juvenile Rhumatoid arthritis, also known as Juvenile Idoipathic arthritis. Fifty thousand children have JRA in the US alone, with many having symptoms into adulthood. While it is more common in girls with Turner syndrome than in the general population, it is not one of the more common symptoms of Turner syndrome, and affects no more than four percent of girls with Turner syndrome.

Bicuspid Aortic Foundation
bicuspidfoundation.com
The foundation provides support and up to date information for those with a bicuspid aortic valve (BAV).

The Canadian Hard of Hearing Association
www.chha.ca
CHHA is a bilingual consumer organization run by and for persons who are hard of hearing. It has branches and chapters in Canada from coast to coast. The national office is in Ottawa. They work to give hard of hearing Canadians a chance to hear and be heard. Anyone who wants to improve conditions for those with a hearing loss are welcome to join.

Hearing Link

hearinglink.org
This UK organization is for people with hearing loss and their families. Their website lists their services and the support they offer, which is very comprehensive. They enable people to connect and share experiences and advice. There are an estimated ten million people in the United Kingdom with hearing loss.

The Hearing Loss Association of America
www.hearingloss.org
This organization which represent and supports people with hearing loss, works to eradicate the stigma associated with hearing loss and raises awareness for the need for prevention, treatment, and regular hearing screenings throughout life. It has more than two hundred local and state chapters across the United States.

The organization also works at the national level on legislation that impacts people with hearing loss. Approximately 17 percent of the American population (more than 36 million people) have hearing loss that impacts their everyday life. They have accomplished many things that Americans now take for granted. Watch the video on their web site.

Little People of America
www.lpaonline.org

Turner syndrome is listed among the 200 reasons for short stature that fall under the umbrella of the Little People of America. The organization holds annual conferences, and their website is an excellent source for adaptive products such as pedal extenders for driving.

RESOLVE: The National Infertility Organization
www.resolve.org

This organization provides support and runs support groups all over the United States for those struggling with infertility.

Appendix 8

Glossary

Glossary of terms mentioned in Standing Tall With Turner Syndrome

* Important note: This glossary is limited to medical and cultural references mentioned in the book, so that readers, no matter their age, country, or background can understand the work. This is not a comprehensive list of all possible issues in Turner syndrome, but is to insure comprehension of the specific subjects that the writers raise.

A

Abzug, Bella: (July 1920-March 1998) She spent her life fighting for social and political change. A leading American liberal activist and politician in the 1960s and 1970s, she was known for her women's rights advocacy. Rejected by Harvard Law School because of her gender, she did not give up, but earned a law degree from Columbia University in 1947.

Amniocentesis: A test also called amniotic fluid test (AFT), done between the fifteenth and twentieth week of pregnancy. A long needle is inserted into the pregnant woman's abdomen about twenty cc's (a bit more than half an ounce) of the fluid surrounding the fetus is taken out and used for karyotyping and other tests to check the health of the baby.

Androgens: Usually thought of as male hormones, (testosterone is one of them). the female body also uses them in smaller amounts. In women they are produced in the adrenal gland, ovaries and fat cells. They are important for growth and the function of many organs. They are even required for the body to use estrogens properly. They stimulate hair growth in the underarm and pubic areas, and also stimulate sex drive.

Anecdotal evidence: Evidence based on casual observation a person might make, or on a few stories - not statistical scientific measurements.

Areolae: (singular is areola) The colored circular area around the nipple of the breast.

Autoimmune disorder: A disease in which the immune system makes the mistake of attacking and destroying healthy tissues in the body instead of the body's enemies (antigens) such as viruses, toxins, cancer cells or blood or tissues from another person. Examples of autoimmune disorders include allergies, celiac disease, rheumatoid arthritis, and type 1 diabetes.

B

Bacterial infection: Bacteria are tiny organisms a few micrometers long. They come in many shapes and live all over us and inside us. Most of them are either good for our health or neutral. Only a few types such as Streptococcus (strep) and Staphylococcus (staph) make us sick.

Bicuspid aortic valve: A valve in the heart that is missing a flap. See Appendix 4 for more information.

Biochemistry: The study of chemicals in living things, and how cells use them.

Bone age: The average age which bones reach at a certain stage of maturity. As children get older the bones change in size and shape, and these changes can be seen and measured on an x-ray. Usually an x-ray of the hand shows the stage of growth, so that doctors can predict how long the child is still likely to grow, and possibly to anticipate final height.

C

Cardiologist: A doctor who specializes in heart issues, diagnosis, and treatment, but is not a surgeon.

Centers for Disease Control and Prevention: This federal agency, located in Atlanta, Georgia, was formed in 1946 to fight malaria. Today it is one of the major components of the Department of Health and Human Services and is the United States premier health promotion, disease prevention and preparedness agency.

Chapin, Harry: (December 1942-July 1981) He was an American singer-songwriter and a dedicated humanitarian who fought to end world hunger. Chapin donated a third of his paid concerts to charitable causes. He used to say that "money is for people" and gave it away.

Chorionic villus sampling: (CVS) A prenatal test used as early as eight weeks into a pregnancy to check for genetic problems. It involves taking a small sample of placenta cells, and can detect the same conditions as amniocentesis.

Chromosome: A piece of DNA that contains many genes.

Coarctation: A condition in which the aorta is pinched or narrow in a specific spot (from Latin meaning to "press together").

Contraceptives: Methods that prevent pregnancy. The birth control pill achieves birth control hormonally rather than with a barrier, such as a condom or diaphragm. These hormones can be used for women with Turner syndrome to provide female hormones in the correct amounts.

Computerized tomography (CT): A series of x-ray views taken from different angles. Using a computer, they produce images that are like cross sections of the body, and produce more information that a regular x-ray. They can be used to visualize any part of the body, not just bones.

Cubitus valgus: It is sometimes called "milkmaid's elbow", because of the wide carrying angle of the elbow. The forearm points away from the midline of the body. It is so common in Turner syndrome, that it can be used to aid in diagnosis. Brigitt Angst's photo essay includes a picture showing an elbow bending in this way.

Curve ball: This expression comes from baseball. It means something surprising and unexpected. This ball is hard to hit because it suddenly veers to the side before heading towards the hitter.

Cystitis: Inflammation of the urethra attributable to different causes. It can be due to an irritation of the area from sex, or sensitivity to soap, or even wearing pants that are too tight! It can also be the result of a bacterial infection, but then it is usually referred to as a urinary tract infection (UTI) and treated with antibiotics. Drinking extra fluids can reduce the bacteria by physically flushing them out. Drinking enough to have a large volume of urine at least once an hour can help avoid the need for antibiotics.

D

Dalai Lama: The 14th Dalai Lama, born in July 1935, is the Buddist spiritual leader of Tibet, currently living in exile in Dharamsala, Northern India.

Dexa scan: (dual-energy x-ray absorptionetry) an x-ray to measure your bone mineral density to check for osteoporosis.

DNA: (Dioxyribonucleic acid) Hereditary material. Almost every cell in your body has the same DNA.

E

EasyPod: One of several pen systems to make it easier to give yourself daily growth hormone shots if you have difficulty with a regular syringe.

Endocrinologist: A doctor who specializes in the problems of hormone imbalances and restoring the balance in the body. These disorders include diabetes, too much thyroid (hyperthyroidism), too little thyroid (hypothyroidism), many growth disorders, Turner syndrome, as well as other disorders with a hormonal component.

Estrogen: These are the primary female sex hormones, produced in the ovaries and in smaller amounts in the liver, breasts, adrenal glands and fat cells.F

Fear Factor: An American sports/dare/stunt reality game show with prizes. There are versions in many countries around the world.

Flat: A UK word for an apartment.

G

Gene: A small unit on a chromosome that holds information to build and maintain a person or other living thing. Not everything in the genes is hereditary. Turner syndrome for example, is a mutation in the genes that is not inherited, unless your mother had Turner syndrome.

Genetics: The science of genes and the variations in organisms.

Geneticist: A biologist who studies the science of genes and heredity.

Genitalia: The external organs of the reproductive system.

H

High blood pressure: (also known as hypertension) The excessive force of the blood pushing against the wall of the arteries as the heart pumps. High blood pressure affects one in five adults in the general population.

Horseshoe shaped kidneys: These kidneys are fused together at the bottom and form a horseshoe shape. This is the most common renal fusion problem in the general population, and occurs about fifteen percent of the time in Turner syndrome.

HRT: (Hormone Replacement Therapy)
HRT consists of the female hormones that women need if their ovaries have failed. It is taken by mouth once a day, or worn on the skin in a patch. Many women who do not have Turner take HRT during menopause to ease unpleasant symptoms during the transition. Current thinking is that HRT should stop for all women at the end of the age when menopause should be complete, usually around age fifty.

Hypothyroidism: Condition in which the thyroid gland does not produce enough thyroid hormone. Low levels of thyroxine can measured in the blood. A synthetic form of this hormone like levothyroxine is prescribed in pill form, taken once a day.

I

Inflammation: Part of the immune response to an irritation of some kind. It does not always mean infection, though infections can cause inflammation. Sometimes inflammation, instead of offering protection and healing to bruised tissues, for example, actually makes things worse and causes damage. This is true in cases of chronic inflammation such as Arthritis.

IVF: (In Vitro Fertilization) A leading reproductive technology for infertile couples, egg and sperm are joined outside a woman's body and then the fertilized egg is implanted in the uterus. Pioneered in 1978, it has been used by millions of women, including those with Turner syndrome.

J

Juvenile rheumatoid arthritis: (JRA, sometimes called juvenile idiopathic arthritis) is a chronic arthritis in children under 16, with pain and swelling in the joints. The cause of JRA is still unknown. It is thought to be an autoimmune disorder, (the immune system attacks healthy tissue). JRA, like Turner syndrome, can cause short stature. There are several types.
In one study, www.ncbi.nlm.nih.gov/pubmed/9706435, of sixty-five girls with Turner syndrome, three had JRA. (four percent). In another, http://www.ncbi.nlm.nih.gov/pubmed/9706435, of fifteen thousand children with JRA, eighteen had Turner syndrome, so the prevalence of JRA in Turner syndrome is greater than would be expected if the two conditions were randomly associated. Overall, JRA is more common in girls than in boys.

K

Karyotype: A complete set of the chromosomes in an individual, which are then arranged in pairs, and by size. The picture of this arrangement is sometimes called a karyogram.

Kitt, Eartha: (January 1927–December-2008) She was an American singer, actress, dancer and star. Known for her 1953 hits: *C'est Si Bon* and her novelty Christmas song *Santa Baby*, she was also Catwoman in the third season of the iconic 1960s *Batman* television show, voiced characters for Disney (winning five Emmy Awards) and was a guest star on the hit TV show *The Simpsons*. She was also a Broadway star and cabaret singer. In the late 1990s she played the Wicked Witch of the West for the North American Touring Company alongside our essayist Wendy Coates, who played a Munchkin in *The Wizard Of Oz*.

L

Labor Day: An American federal holiday celebrated on the first Monday in September. It celebrates workers with a day off, but is also considered the last weekend of summer.

Lymph system: A major part of the body's immune system, this network of organs, nodes, ducts, and vessels make and move lymph from tissues to the bloodstream, including white blood cells.

M

Metabolism: The chemical reactions that keep cells alive in your body, and how they get energy and use nutrition.

Mosaic: Used to describe a karyotype when the woman with Turner syndrome has a combination of cells with some missing the second X and some having it. In recent years it has been decided that all those with Turner syndrome have to have some mosaicism in order to live. A karyotype described as "mosaic" does not describe which symptoms of Turner syndrome are present.

MRI: (Magnetic resonance imaging) This test uses powerful magnets and radio waves to create 3 dimensional pictures of the internal organs, blood vessels, etc. without x-rays.

Munchkins: Natives of the fictional Land of Oz, from the books by L. Frank Baum, and the famous 1939 movie *The Wizard Of Oz*. Munchkins are very short.

Mutation: (genetic) A permanent change in hereditary material that alters the message carried by genes. In Turner syndrome, it is a change in DNA, that happens after conception. No one knows how the mutations in Turner syndrome happen. They occur in all ethnic groups all over the world at the same rate. Some mutations run in families, but Turner syndrome is not among them. Rarely a woman with Turner syndrome is fertile, and she has a fifty percent chance of having a baby with Turner syndrome.

N

Nephrologist: A doctor specializing in the function and problems of the kidneys, and illnesses that affect the kidneys like high blood pressure, diabetes, or kidney stones.

Neuroendocrinology: The study of the relationship between the nervous system and the endocrine system, and how the various cells and hormones interact and communicate.

Nin, Anaïs: (February 1903–January 1977) She was an American author born to Spanish-Cuban parents in France, where she was raised. She lived most of her life in the United States, where she became an author and published journals for more than sixty years, beginning when she was eleven years old.

Non Verbal Learning Disorder ((NVD or NVLD): A condition in which people have trouble processing visual information, causing problems in pattern recognition, math or visual memory. In addition, they can have problems reading and interpreting the meaning of facial expressions, or remembering faces.

O

Ob/GYN: This medical specialty includes the entire female reproductive system, whether pregnant or not. The training includes surgery.

Orgasm: The sudden discharge of accumulated sexual tension, manifested by rhythmic muscle contractions and a feeling of euphoria.

Osteoporosis: Bone weakness due to the loss of bone density (calcium and minerals) resulting in fragility and propensity to bone fractures. The condition is more common in post-menopausal women, but occurs in both sexes. The risk can be reduced with lifestyle changes like exercise .

P

Progesterone: A female hormone produced in the ovaries, in the adrenal gland, and also by the placenta during pregnancy. It is one of five major classes of steroid hormones, which includes estrogens and androgens. When progesterone levels in the blood drop, menstruation happens.

Puberty: The physical changes that gradually make the child's body transform into that of an adult. On average, girls start puberty at age ten or eleven, and complete the process by fifteen or so. This process is measured in five physical development stages called the Tanner scale.

Puddle jumper: Puddle jumper: A slang term for a very small commercial airplane with six to twenty seats. They run from large airport hubs to small regional airports, landing in many rural places. Usually a bit bouncy compared to large planes.

R

Raising Arizona: is an award winning 1987 comedy full of visual gags, unconventional characters, biblical references, pathos and funny dialogue. The story is about a couple that wants children but she is infertile. Because the man has a criminal record they cannot adopt. Local furniture magnate, Nathan Arizona has quintuplets so they decide that he has babies to spare and they kidnap one of them.

Rheumatologist: This doctor specializes in issues involving joints and connective tissues, like tendons and ligaments. Autoimmune diseases and connective tissue disorders are known as rheumatic diseases, and include arthritis.

Rooney, Mickey: (born September 1920) is a famous American actor whose career spans from the 1920s until the 2010s. A superstar as a teenager, he starred in dozens of films. He has two Oscars, was nominated for four more, and has many other performance awards. He played the Wizard for the North American Touring Company alongside our essayist Wendy Coates, who played a munchkin in this stage production of *The Wizard of Oz*. He is mentioned in the 1972 song *Celluloid Heroes* by The Kinks: "If you stomped on Mickey Rooney/ He'd still turn 'round and smile..."

S

Sexuality: A person's expression of erotic experiences and responses, and what form their sexual interests take.

Superbowl Sunday: This is the Championship final game of the American football season. Many Americans who are not typically football fans will watch the game or go to a Superbowl party. It is the second largest day of food consumption in the United States after Thanskgiving Day.

Synthroid: A brand name of Levothyroxine, a synthetic thyroid hormone taken when the thyroid stops producing enough thyroxine on its own.

T

Tanner scale: This is the scale of physical development from child to adolescent to adult, for both sexes, in stages one to five. Stage one means no sexual development and stage five means complete sexual development.

Testosterone: A steroid hormone produced in the testicles of men and the ovaries of women. Men have twenty times more testosterone than women, but both sexes also get testosterone from the adrenal gland. Testosterone supplements for any woman, including those with Turner syndrome, are extremely controversial.

Translational research: This research finds applications for scientific discoveries in medicine and other areas.

Translocation: This is the transfer of a segment of a chromosome to another chromosome or a switch to a different spot on the same chromosome.

Thyroid: Shaped like a butterfly, this is one of the largest endocrine glands. It is located in the neck just under the Adam's Apple, and produces hormones that regulate many other systems in the body.

U

Ureters: The tubes that bring the urine from the kidneys to the bladder.

Urethra: The tube that carries the urine from the bladder via the genitals, out of the body. In women this tube is much shorter than in men, which explains why women are more susceptible to bladder infections. The bacteria have a much shorter trip.

X

X-men: Originally in comic books, there is a successful series of movies about the X-men's adventures. They are a team of mutant superheroes that possess the special powers of the "X-gene", which normal humans lack, but give mutants their abilities.

45, XO: A common way to describe the chromosomal situation when the second X chromosome is missing. The "O" signifies a placeholder for the missing X. 45, XO does not mean that an X was replaced by an "O" and does not represent a zero either. More recently, 45, X is now preferred in genetic circles to describe this genetic profile. A normal cell would be 46, XX.

CPSIA information can be obtained
at www.ICGtesting.com
Printed in the USA
BVHW012359100721
611476BV00011B/398